TACTILE SENSING
AND DISPLAYS

TACTILE SENSING AND DISPLAYS

HAPTIC FEEDBACK FOR MINIMALLY INVASIVE SURGERY AND ROBOTICS

Saeed Sokhanvar

Helbling Precision Engineering Inc., USA

Javad Dargahi

Concordia University, Montreal, Canada

Siamak Najarian

Amirkabir University of Technology, Tehran, Iran

Siamak Arbatani

Concordia University, Montreal, Canada

A John Wiley & Sons, Ltd., Publication

Library of Congress Cataloging-in-Publication Data

Tactile sensing and displays : haptic feedback for minimally invasive surgery
and robotics / Saeed Sokhanvar ... [et al.].
      p. ; cm.
   Includes bibliographical references and index.
   ISBN 978-1-119-97249-5 (hardback)
   I. Sokhanvar, Saeed
   [DNLM: 1. Surgery, Computer-Assisted–methods. 2. Touch Perception.
3. Tactile Feedback. 4. Robotics–instrumentation. 5. Surgical Procedures, Minimally
Invasive–instrumentation. 6. User-Computer Interface. WO 505]
   617.90028'4–dc23
                                                         2012026217

A catalogue record for this book is available from the British Library.

Print ISBN: 9781119972495

Typeset in 10/12.5 Palatino by Laserwords Private Limited, Chennai, India
Printed and bound in Singapore by Markono Print Media Pte Ltd

Contents

Preface

Minimally invasive robotic surgery (MIRS) was initially introduced in 1987 with the first laparoscopic surgery, a cholecystecotomy. Before introduction of surgical robots, numerous laparoscopic procedures had been performed with the development of newer technology in conjunction with increased skills acquired by surgeons. This type of surgery is known as minimally invasive surgery (MIS) because incisions are smaller, with the conferred benefits that include less risk of infection, shorter hospital incarceration, and speedier recuperation. One of the present limitations to MIS, however, is that the equipment requires a surgeon to move the instruments while, at the same time, viewing a video monitor. Furthermore, the surgeon must move in the opposite direction from the target on the monitor to interact with the correct area on the patient in order to achieve a reasonable level of hand–eye coordination, tactile and force feedback, and dexterity. Other current drawbacks of laparoscopic surgery include restricted degrees of motion, increased sensitivity to hand movement and, perhaps most significantly, lack of tactile feedback. Although this latter aspect has been studied by many researchers, no commercial MIS or MIRS with tactile feedback is currently available. One of the main reasons for this is the sheer complexity of such systems. However, with the advent of recent advancements in miniaturization techniques, as well as acceptance of surgical robots by many surgeons and hospitals, it seems that now is the right time for a leap into the next generation of minimally invasive surgical robots augmented with tactile feedback.

The objective of this book is to provide readers with a comprehensive review of the latest advancements in the area of tactile sensing and displays applicable to minimally invasive technology and surgical robots, into which the latest and most innovative haptic feedback features will eventually be incorporated. Readers will not only learn about the latest developments in the area of tactile sensors and displays, but also be presented with some tangible examples of step-by-step development of several different types. Haptics, as we know it today, is a multidisciplinary area including, but not limited to, mechanical, electrical, and control engineering as well as topics in psychophysics. Throughout this book, readers will become acquainted with the different elements and technologies involved in the development of such systems. The regulatory aspects of medical devices, including MIS systems and surgical robots, are also discussed.

This book is organized into 12 chapters. Chapter 1 introduces tactile sensing and display systems. Chapter 2 introduces a wide range of tactile sensing technologies. Chapter 3 discusses the piezoelectric polymer PVDF, which is a fundamental composite of several tactile sensors presented in this book. Chapter 4 details the design and micro-manufacturing steps of an endoscopic force sensor as well as a multi-functional tactile sensor. Chapter 5 provides a study on the force signature of different soft materials held by an endoscopic grasper. Chapter 6 focuses on the hyperelastic finite element modeling of lumps embedded in soft tissues. This model uses the Mooney–Rivlin model to investigate the effect of different lump parameters such as size, depth, and hardness on the output of endoscopic force sensors. Chapter 7 provides a review of tactile display technologies. Chapter 8 introduces an alternative tactile display method called a grayscale graphical softness tactile display. Chapter 9 briefly reviews the current state of MIRS. Chapter 10 deals with teletaction and its involved elements. Chapter 11 discusses the design, implementation, and testing of a closed loop system for a softness sensing display. And, finally, Chapter 12 provides a review of the latest regulatory issues and FDA approval procedures.

The authors are deeply indebted to many people for their help, encouragement, and constructive criticism throughout the compilation of this book.

Saeed Sokhanvar
Javad Dargahi
Siamak Najarian
Siamak Arbatani

About the Authors

Saeed Sokhanvar received his B.Sc. and M.Sc. in Mechanical and Biomechanical Engineering from University of Tehran and Sharif University of Technology in 1990 and 1994, respectively. Then he worked for several years in the area of medical devices. He received his PhD in the area of tactile sensing for surgical robots from Concordia University, Canada. While working on his PhD he received several major awards for academic excellence, such as Postdoctoral Fellowship from the Natural Science and Engineering Research Council of Canada (NSERC), a Precarn Graduate Scholarship, a J.W. O'Brien Graduate Fellowship, and an ASME-The First Annual ASME Quebec Section Scholarship, among many others. He then joined MIT's BioInstrumentation Lab as a senior postdoctoral research fellow and worked on projects such as early diagnosis of diabetes and needle-free injection systems. In 2009 he joined Helbling Precision Engineering, a medical design and development firm, in which he has contributed to research and development of a number of medical devices, including drug delivery systems, and minimally invasive surgical tools. In addition to several patents & patent applications, he has published more than 20 papers in renowned journals and conferences.

 Javad Dargahi serves as a Full-Professor in the Mechanical and Industrial Engineering Department at Concordia University in Montreal, Canada. He received his B.Sc. and M.Sc. degree in Mechanical Engineering from University of Paisley, UK and his Ph.D. degree from Glasgow Caledonian University, UK in the area of "Robotic Tactile Sensing". He was a senior postdoctoral research associate with the Micromachining/Medical Robotics Group at Simon Fraser University, Canada. He worked as an Assistant Professor in the Biomedical Engineering Department at Amirkabir University of Technology, as an Engineer in Pega Medical Company in Montreal and as a full-time lecturer in the Engineering Department at University of New Brunswick. His research interests are design and fabrication of haptic sensors and feedback systems for minimally invasive surgery and robotics, micromachined sensors and actuators, tactile sensors and displays and robotic surgery. In addition to several patents and patent applications, Prof. Dargahi has published over 160 refereed journal and conference papers. He is author of two new books published by McGraw-Hill. One of his books "Artificial Tactile Sensing in Biomedical Engineering" was the runner-up in the Engineering & Technology category of the Professional and Scholarly Excellence Awards, which are known as the "Oscars" of the Association of American Publishers in 2009. His second book "Mechatronics in Medicine" was published in 2011. Dr. Dargahi has been principal reviewer of several major NASA proposals in the area of "Crew health and performance in space exploration mission".

Siamak Najarian serves as the Full-Professor of Biomedical Engineering at Amirkabir University of Technology, Iran. He has completed his Ph.D. in Biomedical Engineering at Oxford University, England and had a postdoctoral position at the same university for one year. His research interests are the applications of artificial tactile sensing (especially in robotic surgery), mechatronics in biological systems, and design of artificial organs. He is the author and translator of 35 books in the field of biomedical engineering, 11 of which are written in English. Prof. Najarian has published more than 200 international journal and conference papers in the field of biomedical engineering along with the two international books in the same field. One of his books entitled "Artificial Tactile Sensing in Biomedical Engineering" achieved the rank of finalist at the 2009 PROSE Awards among all the entries in Engineering and Technology category (published by McGraw-Hill Publication).

Siamak Arbatani received a scholarship to conduct his MSc degree in the Mechanical Engineering Department at Shiraz University. He completed his degree with the rank of 2nd best student in the entire department. He worked for the Concept Software Company in USA for a couple of years. Mr. Arbatani joined as a PhD research associate with Dr. Dargahi's research team in January 2011. His research interest is in the area of haptic feedback in robotic assisted minimally invasive surgery, specifically in the development of state of the art haptic displays.

1

Introduction to Tactile Sensing and Display

1.1 Background

Throughout the ages, humans have become accustomed to the environment by using their five senses: sight (vision), hearing (audition), touch (taction), smell (olfaction), and taste (gustation). Most of us subjectively experience the world through these five dimensions, although only two of these, sight and hearing, have been reliably harnessed for the work of objective scientific observation. For the senses of smell, taste, and touch, however, objective and accurate measurements are still being sought. This chapter will deal mainly with the under-represented sense of touch, which perceives temperature, force, force position, vibration, slip, limb orientation, and pain. The sense of touch confers upon us a haptical experience without which it would be difficult to write, grasp a light object, or to gauge the properties of objects [1]. Given the importance of touch (tactile sensing) in scientific work and daily life, researchers have been striving to understand this sense more thoroughly, with the goal of developing the next generation of tactile-based applications. Though the concept of replaying audio and visual recordings is quite familiar to us, the applications and devices for gathering tactile information and rendering it into a useful form is not, as yet, well understood or characterized.

A conceptual comparison between collecting and displaying information for visual, auditory, and tactile systems is shown in Figure 1.1.

Viewed objectively, touch is perceived when external stimuli interact through physical contact with our mechanoreceptors. Contrary to our other senses, which are localized in the eyes, nose, mouth, and ears, the sense of touch is a whole-body experience that comprises arrays of different nerve types and sensing elements. Our skin is capable of sensing force, the position of applied force, vibration (pulsation), softness, texture, and the viscoelasticity of any object with which it comes into contact. This permits us to determine things about any object we touch, such as mass distribution, fine-form features, temperature, and shape. To some extent, these senses that are felt by the fingers can be simulated by using signals from tactile sensors in order to provide proportional input control to any grasping application [2]. Although touch is a whole-body experience, research on touch-based (haptic) systems focuses primarily on the hand and particularly the fingertips, which contain the greatest number of tactile receptors. Tactile information is

Tactile Sensing and Displays: Haptic Feedback for Minimally Invasive Surgery and Robotics, First Edition.
Saeed Sokhanvar, Javad Dargahi, Siamak Najarian, and Siamak Arbatani.
© 2013 John Wiley & Sons, Ltd. Published 2013 by John Wiley & Sons, Ltd.

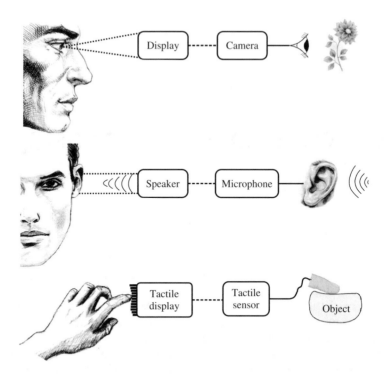

Figure 1.1   Collecting and displaying visual, auditory, and tactile information

gathered by stroking the fingers across an object to provide information about its texture, or by pressing on an object in order to determine how soft or hard it is, or moving fingers around the perimeter of an object to gather information about its shape [3]. Generally, the ways in which the human hand and fingers gather tactile information have been duplicated by researchers when developing touch sensors for similar purposes.

In the 1990s, efforts by researchers to design a commercially viable robotic hand that contained touch sensors proved unsuccessful. This failure was attributed to the sheer complexity of such systems since touch sensors need to physically interact with objects, whereas audio or visual systems do not. Also, tactile sensing may often not be the most effective option in such a highly structured environment as the automated car industry. Nevertheless, for unstructured environments where irregularities occur in any object that is handled, or if there is any disorder in the working environment, the role of tactile sensing in gathering tactile information through haptic exploration is pivotal [4]. It is also evident that the use of remote tactile sensors is preferable in any hazardous or life-threatening environment, such as beneath the ocean or outer space, and for which no other sensing modality, such as hearing or vision, can be substituted. The purpose of this book is to explore some of the features, challenges, and advancements of research in tactile sensing and displays in a number of ongoing research projects in the areas of minimally invasive surgery (MIS) and robotic minimally invasive surgery (RMIS), with the emphasis on novel tactile sensing and display methods.

1.2 Conventional and Modern Surgical Techniques

Surgery is the treatment of diseases or other ailments through manual intervention using instruments that cut and sew body tissues. In open surgery, referred to as the first-generation technique, a large incision was usually made in the body that allowed the surgeon full access to organs and tissues. Although this type of procedure allowed the surgeon to have a wide range of motion, as well as tissue assessment through palpation, the resulting trauma highlighted the limitations of this technique [5]. For instance, in a conventional open-heart cardiac operation, or a cholecystectomy (gall bladder removal surgery), the majority of trauma to the patient is caused by the surgeon's incisions to gain access to the surgical site, rather than the procedure itself. The invasiveness of cracking and splitting the rib cage to uncover the heart muscle, and trauma caused by incisions in the abdomen to gain access to the gall bladder, causes a long postoperative hospital stay and increasing cost and pain to the patient [6].

However, recent advances in surgery have greatly reduced the invasiveness of previous surgical procedures and, to overcome many of the shortcomings and complications of open surgery, MIS (referred to as the second-generation surgical procedure) was introduced. It involves making very small incisions (referred to as access holes or trocar ports) in the body through which very slender devices are inserted. These can be a laparoscopic tool, an endoscope to allow the surgeon full view of the surgical site or a sharp pointed instrument (trocar) enclosed in a metal tube (cannula) either to draw off fluid or to introduce medication. Various other surgical instruments, such as clippers, scissors, graspers, shears, cauterizers, dissectors, and irrigators are also used. These are mounted on long poles and can be inserted and removed from the other trocar ports to allow the surgeon to perform other necessary tasks. Upon completion of the surgical procedure, the trocars are removed and the incisions closed.

There are, however, certain disadvantages with MIS. Because these procedures are viewed on a 2D screen, the natural hand–eye coordination is disrupted, as indeed is the surgeon's perception of depth [7]. Furthermore, because images from the camera are magnified, small motions such as tremors in the camera, or even a heartbeat, can cause the surgical team to experience motion-induced nausea. In addition, because there is no tactile feedback, the surgeon has no sense of how hard he is pulling, cutting, twisting, or suturing.

There is a phenomenon referred to as SAID (specific adaption to imposed demands), which can loosely be interpreted as meaning that your body always adapts to exactly what you are doing, whether you are conscious of it or not. In terms of MIS, this means that the surgeon is required to undergo a rigorous retraining program, otherwise he/she is limited to performing either restricted or simpler surgical procedures. Because modern surgical procedures are far more complex than in the past, and the surgeon's knowledge and skill alone may not guarantee the success of an operation [8], a third-generation of surgical procedures, robotic surgery, was developed to rectify this potential problem since it places less stress on the surgeon and decreases both the surgical time and the patient recovery period [9, 10]. However, the current medical robots have not eliminated all the above-mentioned shortcomings. As an example, although the Da Vinci™ and Zeus medical robots have rectified a number of them, surgeons are still handicapped by the lack of tactile sensing and its associated benefits [11, 12].

The ability to distinguish between different types of tissue in the body is of vital importance to a surgeon. Before making an incision into tissue, the surgeon must first identify what type of tissue is being incised, such as fatty, muscular, vascular, or nerve tissue, since failure to classify this correctly can have severe repercussions. For example, if a surgeon fails to properly classify a nerve and cuts it, the patient may suffer effects ranging from loss of feeling to loss of motor control. The identification and classification of different types of tissue during surgery, and more importantly during a cutting procedure, will necessitate the creation of smart surgical tools. One major approach to developing such tools is retrofitting existing surgical tools, instead of developing new and revolutionary ones. The advantage of this is due to the fact that newly designed surgical tools are subject to the strictures of regulatory control bodies, such as the Food and Drug Administration (FDA) and the European Community (EC), under whose auspices approval could take anywhere between 5 and 15 years [13]. Therefore, retrofitting current surgical tools is preferable and, in any event, these modified tools have already been used by surgeons and who are, therefore, familiar with their applications. Furthermore, the cost involved in clinical trials for such retrofitted tools can be either avoided altogether or at least greatly reduced.

In conclusion, smart endoscopic tools are a prerequisite to enhancing and facilitating the performance of currently available MIS procedures and robotic surgery. Further development of MIS tools is required, as is the necessity to expand upon our existing knowledge and research into the industrial aspects of robotic surgery.

1.3 Motivation

Minimally invasive procedures are growing rapidly. Presently, 40% of all surgeries are performed in this manner and this will rise to 80% by end of the decade [14]. However, despite the many advantages of MIS and MIRS (minimally invasive robotic surgery) over conventional surgery, their main drawback is the almost complete lack of a sense of touch. Even the only FDA-approved robots (Da Vinci™ from Intuitive Surgical Inc. and Zeus from Computer Motion Inc.) suffer from this lack of haptic feedback and the latest and most modern system, the Amadeus Robotic Surgical System from Titan Medical Inc. (which is planned for release in 2012) benefits from only force feedback capabilities. Teletaction systems generally comprise three key elements, namely: a tactile sensor array, a tactile filter (processing, conversion, and control algorithms), and a tactile display. The tactile sensor array collects comprehensive information about the contact zone. This information is then processed using a tactile filter and the control signals are then passed to a tactile display which is able to mimic softness, roughness, and texture in a static and dynamic way. Therefore, restoring the lost tactile perception has been the motivating factor in several recent research works, including this book. In addition to the introduction of an innovative multifunctional tactile sensor that can be integrated into conventional MIS tools, exploring the potential capabilities of such sensors in terms of force measurement, force position sensing, and softness sensing is of high importance. Another challenging problem in this area is the development of methodologies for processing and converting the data gathered by tactile sensors into a useful format for surgeons. In this regard, several attempts have been made to develop different kinds of tactile displays. The complexity of the mechanical tactile displays provided the motivation for introducing a scheme in which the tactile information is graphically presented to the surgeon.

1.4   Tactile Sensing

Tactile sensing can be defined as a form of sensing that can measure given properties of an object through physical contact between the sensory organ and the object. Tactile sensors, therefore, are used for measuring the parameters of a contact between the sensor and an object, and so are able to detect and measure the spatial distribution of forces on any given sensory area, including slip and touch sensing.

Slip, in effect, is the measurement and detection of the movement of an object relative to the sensor. Touch sensing can be correlated with the detection and measurement of a contact force at a specified point. The spectrum of stimuli that can be covered by tactile sensing ranges from providing information about the status of contact, such as presence or absence of an object in contact with the sensor, to a thorough mapping or imaging of the tactile state and the object surface texture [15].

There are two determining factors in the design of tactile sensors: the first is the type of application and the second is the type of object to be contacted [16]. For example, unlike hard objects, when the tactile sensor is targeted toward soft objects (e.g., most biological tissues), more complexities arise and there is a need for more sophisticated designs. In general, tactile sensors can be divided into the following categories: mechanical (binary touch mechanism), capacitive, magnetic, optical, piezoelectric, piezoresistive, and silicon-based (MEMS or micro-electro-mechanical systems).

One of the most interesting and relatively new application areas for the tactile sensor is in robotics, MIS, and MIRS [17]. Providing touch to a robot allows it to manipulate delicate objects and to assess their shape, hardness, and texture. Some of these robots, which are already being used in medical surgery, possess haptic capabilities to 'feel' organs and tissues and then transmit this information to the surgeon via an instrument–patient interface, thus replacing the human sense of touch.

1.5   Force Sensing

Force sensing is a basic and necessary capability of tactile sensors that has been investigated for a long time. Nowadays, force sensors of advanced design, for both concentrated and distributed force/pressure measurement, are available. The majority of tactile sensors work on piezoelectric, piezoresistive, and capacitive techniques, or a combination of these properties [16, 18–20].

1.6   Force Position

The capability of finding the position of an applied load is believed to be very useful in MIS procedures. A homogeneous soft object compressed between two jaws of a MIS grasper experiences a smooth distributed load. However, as shown in Chapter 4, the presence of an embedded lump in a grasped soft object appears as a point load superimposed on the distributed background load. Therefore, one of the immediate and most interesting applications of force position sensitivity, as shown in this book, is in locating any hidden features in a bulky soft object. Nevertheless, the application of force position sensitivity is not only limited to lump detection.

1.7 Softness Sensing

The softness/hardness of objects is defined as the resistance of its material to deformation (or indentation) [21, 22]. Hardness sensing is already in use in industry and there are currently specific procedures to measure the hardness of objects. The softness of objects is most commonly measured by the Shore (durometer) test. This method measures the resistance of the object toward indentation and provides an empirical hardness number that does not have an explicit relation to other properties or fundamental characteristics. The Shore hardness using the Shore A, D, or OO scale is the preferred method for rubbers and elastomers. While the Shore OO scale is used to measure the softness of very soft materials, the Shore A scale is used for soft rubbers, and the Shore D scale is used for harder ones. The Shore A softness is the relative softness of elastic materials such as rubber or soft plastics and can be determined with an instrument called a Shore A durometer [21]. International Rubber Hardness Degrees (IRHD) also introduces a measurement scale for this purpose. Although several researchers have attempted to measure the softness of objects in different ways, it is usually Young's modulus that is used to relate the softness of objects by a nonlinear relationship which represents how much spring force a rubber component will exert when subjected to deformation.

However, in order to measure the softness of tissues, one must also consider the behavior of the contact object itself. Since soft tissues are nonlinear and are comprised of viscoelastic materials, they show hysteresis in loading/unloading cycles. Furthermore, variation in characteristics of the different soft tissues adds even more complexity to the problem. Characterization of the soft tissues has also been restricted by the fact that the behavior of the soft tissues differs between *in vivo* and *ex vivo* conditions.

One method is to differentiate between the natural frequency shift of the piezoelectric material and the contact object which, in the case of wood and silicone gum, for example, is about 750 Hz [23]. Yamamoto and Kawai [24] used a rotational step motor to create a screw-like motion in soft tissue and then measured the transient response resulting from these mechanical torsional steps. At the moment, the viscoelasticity of the epidermis is evaluated by analyzing the voltage waveform of the step-motor inducting coil. This waveform is characterized by overshoot, damping ratio and undamped natural frequency. Hardness evaluation was carried out by Bajcsy [25] who pressed a robotic finger, fitted with a low spatial resolution tactile sensor, against an object. This loading and subsequent unloading process was performed in small incremental displacement steps and the sensor output reading on each occasion was recorded.

Material hardness was ranked according to the slopes of the linear parts of the loading and unloading sensor outputs. Work along similar lines was reported by Dario *et al*. [26], using a single element sensor made of a piezoelectric polymer pressed against flat sheets of rubbery materials of different compliance and backed by a reference load cell.

Hardness ranking was associated with the slope of the straight line obtained in the sensor output reference cell signal plane under loading. De Rossi *et al*. [27] proposed the use of charged polymer hydrogels as materials useful in tactile sensing, in particular for softness perception, because of their ideal compliance matching with human skin. Softness sensing has been applied as a diagnostic tool, such as in the case of diabetic neuropathic subjects in which the hardness of foot-sole soft tissue increases in different foot-sole areas [28]. The use of an active palpation sensor for detecting prostate cancer and hypertrophy is reported by Tanaka *et al*. [29].

Although a number of tactile sensors have already been designed, analyzed, and manufactured [15, 29–32], some of them for MIS applications, most are confined to force sensing. In addition, the proposed tactile sensors are either very complex in structure and operation, or difficult to fabricate on a micro-scale. For instance, Shikida *et al*. [32], produced a tactile sensor that consists of a diaphragm with a mesa at the center, a piezoresistance displacement sensor at the periphery, and a chamber for pneumatic actuation that is able to detect both the contact force and hardness of an object. To detect the hardness distribution, the contacted mesa element is pneumatically driven toward the object, whereupon the contacted region of the object is deformed according to the driving force of the mesa element and the hardness of the object. Then, from the relationship between the resultant deformation and the driving force generated by pneumatic pressure, the hardness of the contact object can be evaluated. Furthermore, the proposed tactile sensor can be miniaturized, which makes it attractive, although arranging to have a precisely controlled pneumatic drive, in conjunction with a matrix of indicated sensors, could prove to be a formidable task. Dargahi [15] proposed a softness sensor which consists of two coaxial cylinders. The outer cylinder is pliable and the inner one is rigid. The cylinders are attached to a base plate in such a way that two circular polyvinylidene fluoride (PVDF) films are also inserted between the cylinders and base plate. When the tactile sensor is in contact with an object, depending on the softness of the object, the ratio of the PVDF films' outputs will be different. Although the proposed idea of softness measurement is original, it is difficult to fabricate the sensor through conventional micro-machining procedures, due to the use of flexible material. Also, in order to change the working range of the sensor to one within which the Young's modulus for all objects can still be measured, the Young's modulus of the pliable cylinder must also be changed, which introduces further challenges.

Another approach by Lindahl *et al*. [33] to detect the physical properties, stiffness, and elasticity of human skin, utilized a piezoelectric sensor which oscillated at a resonant frequency. Other parts of this sensor included an electronic vibration pick-up device and a PC equipped with software for measuring the change in frequency when the sensor was attached to an object. This sensor was essentially developed for use as a hand-held device and there is no report on sensor micro-fabrication. In addition, no experiments associated with MIS applications were conducted, so, apart from its capability to detect an embedded lump within tissue, no examination was made to determine its capacity to measure load and tissue softness.

1.8 Lump Detection

Currently, palpation is not possible using MIS, even though it is of special importance and is routinely used by surgeons in open surgery to distinguish between normal and abnormal tissues. It is also used to detect the composition of biological tissues and the consistencies of tumors, which, depending upon the disease, often vary within specific tissues [34]. Although the localization of hidden anatomical features has been the subject of much research [35–39], these studies have largely focused on breast cancers, hence their findings cannot be directly used for MIS applications. Kattavenos *et al*. [40] reported the development of a tactile sensor for recording data when the sensor is swept over a phantom sample containing simulated tumors. However, it was not possible to extract any data from this article regarding the size and depth of any lump that was ostensibly found.

1.9   Tactile Sensing in Humans

As previously mentioned, touch is one of the five human senses, and, although all of the senses are important, touch is often subordinated as being one of the lesser. This is perhaps not surprising given the fact that sight and hearing are usually considered as being the two most important in our everyday lives. Although the senses of smell and taste have their deserved place in our cognitive senses, it is the sense of touch that has always been very important to our existence because it has enabled us to use tools with great dexterity. It therefore follows that a paramount step in the realm of tactile sensing would be to develop a mechanical device that would duplicate and replace the sense of touch in the human hand. Designing hardware for this purpose, however, poses a significant challenge, primarily because our hands (especially the fingers) are extremely sensitive to even the smallest vibrations over a range from 10 to 100 Hz [41]. To help define the design requirements of such hardware, it is important to understand the fundamentals of tactile sensing mechanisms, since this knowledge will greatly assist in improving current designs. In addition, it will aid our understanding of which signals are important to communicate and the extent to which they should be communicated. Although this knowledge is a prerequisite, there is more to consider than simply the neurophysiology of touch, which is, in reality, a combination of tactile and kinesthetic information. The combination of cutaneous and kinesthetic sensing is referred to as haptic perception and it is this, in addition to the interpretation of these signals at a higher level, which ultimately interests us.

1.10   Haptic Sense

Two groups of physiological sensations are involved in the wider meaning of haptic sense. The first is the tactile sense where the receptors are located under the dermis and are known as mechanoreceptors. Tactile sense detects the information from the skin surface, such as contact pressure or vibration. The other is the proprioceptive sense, where the receptors exist in the muscle or in the tendon.

1.10.1   Mechanoreception

The human hand contains a complex array of specialized receptors that are rugged enough to survive repeated impacts, while still retaining the ability to detect faint vibrations and the softest touch.

Four main types of tactile mechanoreceptors have thus far been identified [42, 43], of which each is associated with a specific phenomenon: pressure, shear, vibration, and texture [43]. The sensing element of each of these mechanoreceptors is very similar in that they possess physical packaging and position within the skin that is exclusively adapted to its purpose. Figure 1.2 shows a cross-sectional view of the skin on a human fingertip and the position of specialized touch receptors underneath the skin surface [44, 45]. The distribution of the mechanoreceptors on the palm is shown in Figure 1.3 [46–48].

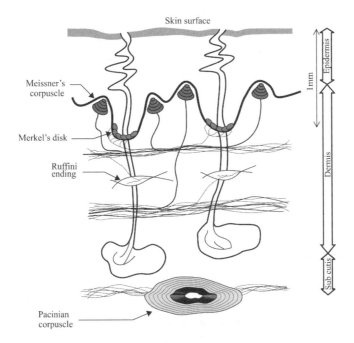

Figure 1.2 Mechanoreceptors [44, 45]

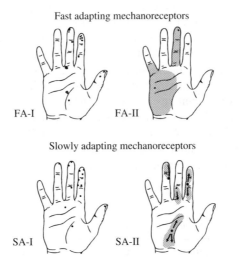

Figure 1.3 Distribution of the mechanoreceptors

Mechanoreceptor types are divided into two categories based on their placement beneath the surface of the skin [45, 49]. Type I receptors (Merkel and Meissner) are located near the surface of the skin between the epidermis and the dermis on the papillary ridges. Type II receptors (Ruffini and Pacinian) are located deeper beneath the skin in the dermis.

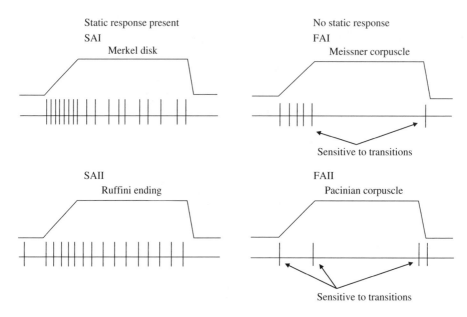

Figure 1.4 Responses of the four types of mechanoreceptors to normal indentation of the skin [45, 47, 48]

Receptors that lie deeper beneath the skin have larger receptive fields. Correspondingly, fewer type II receptors are observed per unit area of skin. Receptors are further divided into fast-adapting (FA) types and slow-adapting (SA) types. FA types do not respond to static stimulus, but only to skin indentation while the stimulus is changing. SA types exhibit a sustained discharge while a steady indentation is maintained. These two types are analogous to the difference between piezoelectric and piezoresistive sensing elements, respectively [48]. However, unlike man-made piezoelectric and piezoresistive sensors, which provide analog signals, biological mechanoreceptors encode their signals as a series of pulses, as shown in Figure 1.4, which are similar to digital serial communication.

Table 1.1 summarizes the characteristics of each mechanoreceptor found in the human fingertip skin and the physical parameters they measure [50].

Table 1.1 Mechanoreceptor characteristics [50]

Receptor	Receptor type	Field diameter (mm)	Frequency range (Hz)	Sensed parameter
Merkel disks	SAI	3–4	DC-30	Local skin curvature
Ruffini endings	SAII	>10	DC-15	Directional skin stretch
Meissner corpuscles	FAI	3–4	10–60	Skin stretch
Pacinian corpuscles	FAII	>20	50–1000	Unlocalized vibration

1.10.2 *Proprioceptive Sense*

Proprioception (from Latin proprius) means 'one's own' and refers to perception of proprioceptive senses. Proprioceptive sense provides the inner information of the body, such as joint angle or muscle contractile force and it keeps track of primarily internal information about body position and movement through the combination of inputs from both the kinesthetic and vestibular senses [51–54]. Kinesthesia informs about the position of body parts with respect to each other, while the vestibular sense details the position of the body part in the world through sensing gravity and acceleration.

1.11 Tactile Display Requirements

The main challenges to the ideas of remote manipulation are those that seek to confer, to the operator, the same sense of touch and proprioception as is felt by the hand (end effector) on a robot manipulator (i.e., temperature, pressure distribution, vibrations, contact geometry). Nowadays, remote manipulation systems provide a comprehensive visual and handling force feedback to the operators. The latest research in this area show that the greatest haptic feedback for the user is provided by combining tactile and kinesthetic feedbacks.

According to the high sensitivity of the human tactile sense at the fingertips, tactile display elements must match different requirements, concerning, for example, the distance of the stimulator elements (spatial resolution), their displacement, temporal resolution, and exerted forces [55]. The most important capabilities of the tactile sense, based on psycho-physiological investigations, are summarized below. The two-point discrimination threshold describes the minimum distance of two stimulation points that a human can still distinguish.

Figure 1.5 shows the two-point discrimination threshold at different regions of the hand, given in [56]. Its value at the fingertips is between 1 mm [57] and 2 mm [58] for vibrational stimulation. For stimulation of the whole area of an average-sized hand, including the palm

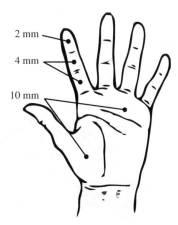

Figure 1.5   Values for the two-point discrimination threshold at different regions of the hand [56]

and the fingers, a tactile display needs to have more than 1000 stimulator elements. The maximum frequency of perceptible vibrations is 1000 Hz [59], in which the maximum sensitivity is at 250 Hz [58]. The maximum deformation of the skin at the fingertips is 3–5 mm [60]. The pain threshold is 3.2 N at a pin diameter of 1.75 mm, corresponds to a pressure of 1.3 MPa [60]. The minimum perceptible deformation is between 0.07 mm [56], when the finger is allowed to move across a pattern and 2 mm [56] without relative motion. Moy and Singh [57] specified the requirements for an ideal device for realistic tactile feedback as being 500 mN mm^{-2} peak pressure, 4 mm stroke, 50 Hz bandwidth and an actuator density of 1 per mm^2.

A realistic feedback in haptic devices, and one which is an important requirement in tactile displays, is that which combines both kinesthetic and tactile feedback. The tactile sensation in real manipulation tasks depends on the current position of the human fingers and is always accompanied by kinesthetic sensations. For kinesthetic feedback, force feedback gloves exist that measure the finger positions for calculating the appropriate forces. A tactile display incorporated into a data/force feedback glove would complete the haptic sensation. This demands a thin, flexible, and lightweight tactile display that allows free movement of the hand and fingers.

1.12   Minimally Invasive Surgery (MIS)

As previously mentioned, MIS is the practice of performing surgery through small incisions or 'ports' in the body through which specialized surgical instruments are inserted, as shown in Figure 1.6. During conventional 'open' surgery, significant trauma is created at the incision site, which results in postoperational pain and discomfort [61–63]. By contrast, MIS procedures result in reduced bleeding and discomfort, improved patient recovery time, and reduced cost.

Notwithstanding its theoretical advantages, however, MIS is more difficult to perform than conventional open surgery due to the lack of advanced tools and the extensive training

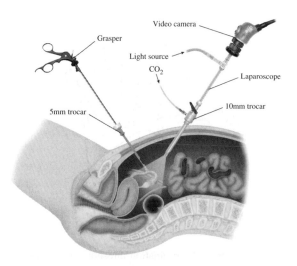

Figure 1.6   Access to internal tissues during surgery

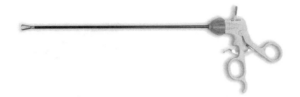

Figure 1.7 Commonly used needle driver for minimally invasive surgery

requirements to familiarize surgeons as to the full potential of this technique. Therefore, MIS is limited to a number of relatively simple procedures, such as cholecystectomy [64] (gall bladder removal) for which there is actually a consensus within the medical fraternity that such practice is, in fact, beneficial. Perhaps the most common form of MIS is laparoscopy [65–67] during which the patient's abdomen is insufflated with CO_2, and cannulas (essentially metal tubes) in which pneumatic check valves are passed through small (approximately 1–2 cm) incisions to provide entry ports for laparoscopic surgical instruments. The instruments include an endoscope for viewing the surgical site (a CCD camera/lens combination with a slender shaft), and tools, such as needle drivers, graspers, scissors, clamps, staplers, and electrocauteries. The instruments differ from conventional instruments in that the working end is separated from its handle by an approximately 30 cm long, 4–13 mm diameter shaft (Figure 1.7) [68].

The surgeon passes these instruments through the cannula and manipulates them inside the abdomen by sliding them in and out, rotating them about their long axis and pivoting them about centers of rotation defined roughly by their incision site in the abdominal wall. Typically, a one-degree-of-freedom device (gripper, scissors, etc.) can be actuated with a handle via a tension rod running the length of the instrument. As shown in Figure 1.8, the surgeon monitors the procedure by means of a television monitor which displays the abdominal work-site image provided by the laparoscopic camera.

1.12.1 Advantages/Disadvantages of MIS

[61, 62, 70–72]:

1. Visualization of the surgical site is reduced. The operating site is viewed on an upright, two-dimensional video monitor placed somewhere in the operating room. The surgeon is deprived of three-dimensional depth cues and must learn the appropriate geometric transformations to properly correlate hand motions to the tool tip motions.
2. The surgeon's ability to orient the instrument tip is reduced. The incision point/cannula restricts the motions of the instrument from six DOF to four. As a result, the surgeon can no longer approach tissue from an arbitrary angle and is often forced to use secondary instruments to manipulate the tissue in order to access it properly or to use additional incision sites. Suturing becomes particularly difficult.
3. The surgeon's ability to feel the instrument/tissue interaction is virtually eliminated. The instruments are somewhat constrained from rotating and sliding within the cannula due to sliding friction with the air seal, and the body wall constrains pivoting motions of the instrument shaft. The mechanical advantage designed into MIS instruments reduces the ability to feel grasping/cutting forces at the handle [72–76].

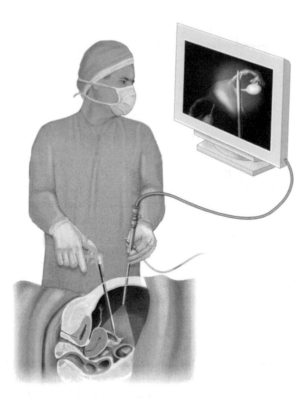

Figure 1.8   The surgeon monitors the procedure by means of a television monitor [69]

Despite the surgeon's considerable skill and ability to work within the constraints of the current MIS technology, the expansion of minimally invasive medical practice remains limited by the lack of dexterity with which surgeons can operate while using current MIS instruments.

1.13   Robotics

The field of robotics may be more practically defined as being the study, design, and use of robot systems for manufacturing purposes. As defined by ISO, an industrial robot is an automatically controlled, reprogrammable, multipurpose manipulator programmable in three or more axes. Typical applications of industrial robots include welding, painting, assembly, pick and place, packaging and palletizing, product inspection, and testing, all accomplished with never-ending endurance, speed, and precision. In the context of general robotics, most types of robots would fall into the category of robotic arms (inherent in the use of the word manipulator in the above-mentioned ISO standard). Degrees of autonomy for robots are variable. Some robots are programmed to carry out specific actions repeatedly with a high degree of accuracy. There are other kinds of robots, however, which are smarter and more flexible as to the orientation of the object on which they are operating, which the robot may even need to identify. Machine vision, in addition to artificial intelligence, is becoming an increasingly important factor in modern industrial robots.

In robot-assisted surgery, which was initially developed to overcome the limitations of MIS, the surgeon uses a computerized control console to manipulate instruments attached to robot arms. This console translates the surgeons' hand and finger movements, scales movements, filters tremors and carries out the operation on the patient in real-time. It is now even possible for specialists to perform remote surgery on patients and for robots to undertake automated (unmanned) operations.

Some advantages of robotic surgery, under optimal and proven conditions, are its greater precision, miniaturization, smaller incisions, decreased blood loss, less pain, quicker healing time, articulation beyond normal manipulation, and three-dimensional magnification, resulting in improved ergonomics. Robotic techniques are also associated with reduced hospital confinement, blood loss, transfusions, and use of pain medication.

Nowadays, there are many applications for robotic-assisted surgery, such as general surgery, cardiothoracic surgery, cardiology and electrophysiology, gastrointestinal surgery, gynecology, neurosurgery, orthopedics, pediatrics, radio surgery, urology, and vascular surgery. Currently, during robotic-assisted surgery, surgeons use magnetic resonance imaging (MRI) or computed tomography (CT) to identify softness discontinuities of tissues preoperatively; however, these imaging techniques do not provide haptic feedback, which makes it very difficult for surgeons to map precisely any preoperative images with intraoperative situations. Recent experimental tests have proved that the presence of direct force feedback significantly reduces the force applied by the Da Vinci™ robot graspers to the tissue [77].

The disadvantages of robotic-assisted surgery are the high cost of the robot itself, the disposable supply cost for each procedure and the intensive learning and training period needed to operate the system.

The ideal goals for the robot industry are to achieve self-maintenance, autonomous learning, and to avoid any harmful situation to people, property, and the robot itself, as well as safe interaction with human beings and the environment. In an attempt to achieve these goals, an ambitious new development is the advent of humanoid robots, which are recurrently being used as a research tool in several scientific areas. In this field, researchers are required to understand the human body structure and behavior (biomechanics) in order to build and test these entities and it is hoped that this attempt to replicate the human frame, together with all its dynamics, will lead to a better understanding of the human body. Like other mechanical robots, a humanoid suggests use of basic components, such as sensing, actuating and planning, and control mechanisms. Since these robots purport to simulate the structure and behavior of a human, and because they are autonomous systems, it is inevitable that humanoid robots are more complex than other kinds of robots. The purpose of their creation is to imitate some of the same physical and mental tasks that humans undergo daily. Scientists and specialists from many different fields, including engineering, cognitive science, and linguistics have combined their efforts to create a robot as human-like as possible. Their ultimate goal is that one day such a robot will have the same deportment, intelligence, and reasoning capability as a human. With such an advent, it is evident that these robots could eventually work in cohesion with humans to create a more productive and brighter future. Another important benefit of developing humanoids is to obtain a greater understanding of the human body's biological and mental processes, from the seemingly simple act of walking, to gaining awareness of the concepts of consciousness and even spirituality.

Figure 1.9   Shadow Dextrous Hand holding a light bulb [78] (© Shadow Robot Company 2008)

Sensing plays an important and prime role in robotic paradigms, the other two being planning and control. Proprioceptive sensors sense the position, orientation and speed of the humanoid's body and joints, and accelerometers measure velocity using integration; tilt sensors measure inclination, force sensors affixed in the robot's hands and feet measure what contact forces have taken place with the environment; position sensors show the actual position of the robot (from which the velocity can be calculated by derivation or even speed sensors). Arrays of tactile sensors can be used to provide data on what has been touched. Tactile sensors will provide information about those forces and torques that have been transferred between the robot and other objects. As shown in Figure 1.9, Shadow Dextrous Hand [78] reproduces all the movements of the human hand and provides comparable force output and sensitivity that makes it possible to manipulate delicate objects such as fruit and eggs. For this same reason, and because it is much weaker than a person, it is safe to use in the presence of human beings and is currently being used for telepresence applications, rehabilitation, and assistive technology, as well as for ergonomic research.

The human body does, however, have much greater flexibility and freedom of movement than can currently be transferred into a humanoid form and so can perform a much greater range of tasks. Although these characteristics would be desirable in humanoid robots, the logistics and complexities required are not presently considered either feasible or viable.

1.13.1   Robotic Surgery

Robotic surgery, or computer-assisted surgery, are terms for technological developments that use robotic systems to aid in surgical procedures. Robotic surgery was developed to overcome both the limitations of MIS and to enhance the capabilities of surgeons performing open surgery. In the case of MIRS, instead of directly moving the instruments, the surgeon uses one of two methods to control the instruments; either a direct telemanipulator or a computer control. A telemanipulator is a remote manipulator that allows the surgeon to perform normal movements that he would encounter during a regular surgical procedure. In computer-controlled systems the surgeon uses a computer to control the robotic arms and its end-effectors, though these systems can also still use telemanipulators for their input. One advantage of using the computerized method is that the surgeon does not have to be present and, indeed, the surgeon could be located anywhere, which ushers in the possibility for remote surgery. In the case of enhanced open surgery, autonomous instruments (in familiar configurations) replace traditional steel tools, performing certain actions (such as rib spreading) with much smoother, feedback-controlled motions than could ever be achieved by a human hand. The main object of such smart instruments is to reduce or eliminate the tissue trauma traditionally associated with open surgery, without requiring more than a few minutes' training on the part of surgeon. This approach seeks to improve that lion's share of surgeries, particularly cardio-thoracic, that minimally invasive techniques have so far failed to supplant.

1.14   Applications

Artificial organs, like an artificial finger or artificial skin, can receive benefits from the use of tactile sensors. For instance, the Shadow Finger Test Unit is an artificial finger which is equipped with tactile sensors such as is used on Shadow Dextrous Hand mentioned above (Figure 1.9). Synthetic skin, fabricated from semiconductor materials that can sense touch, is anticipated to augment robotics in conducting rudimentary tasks that would be considered delicate and require 'touch'. It is also expected that this technology can be further advanced such that it can be eventually used on prosthetic limbs to restore the sense of touch. Markets with distant palpable products may also have a future in tactile sensing applications. Tactile sensors, which have been developed specifically for use with industrial robots, can also complement visual systems by becoming the controlling system during the time in which contact is made between a gripper of the robot, and the object or objects being gripped, since it is during this time that vision is often obscured.

In robotics, tactile sensors provide useful information about the state of contact between a robot hand and the object it is grasping. Sensors can indicate the presence or shape of an object, its location in the hand, and the force of contact.

As previously stated, the lack of haptic feedback is evident in MIS, remote surgery and remote manipulation tasks. Therefore, the significant challenge in the fields of tactile sensing and displays (teletaction systems) is to provide haptic feedback in order to increase operating dexterity and to encourage the greater use of MIS and MIRS.

In robotic telemanipulators, specifically in MIS and MIRS, and in order to provide local shape information, an array of force generators are used to create a pressure distribution on a fingertip that is tantamount to providing a true contact [79]. A comparatively large number of devices have already been developed, and some are commercially available for medical applications. A remote palpation system has been developed that conveys tactile information from inside a patient's body to the surgeon's fingertips during minimally invasive procedures [80]. This system contains tactile sensors that measure pressure distribution on the instruments while tissue is being manipulated. The signal from the sensors is then sampled by a dedicated computer system which applies appropriate signal processing algorithms. Finally, the tactile information is conveyed to the surgeon through tactile display devices that recreate the remote pressure distribution on the surgeon's fingertips. Creation of remote palpation technology will increase safety and reliability in current minimal invasive procedures, and bring the advantages of these techniques to other, more complex procedures, which are not yet available.

References

1. Robles-De-La-Torre, G. (2006) The importance of the sense of touch in virtual and real environments. *IEEE Multimedia*, **13**, 24–30.
2. Gentilucci, M., Toni, I., Daprati, E., and Gangitano, M. (1997) Tactile input of the hand and the control of reaching to grasp movements. *Experimental Brain Research*, **114**, 130–137.
3. Lederman, S.J. and Klatzky, R.L. (1993) Extracting object properties through haptic exploration. *Acta Psychologica*, **84**, 29–40.
4. Lee, M.H. (2000) Tactile sensing: new directions, new challenges. *The International Journal of Robotics Research*, **19**, 636–643.
5. Bicchi, A., Canepa, G., Rossi, D.D. *et al.* (1996) A sensorized minimally invasive surgery tool for detecting tissue elastic properties. Proceedings of the 1999 IEEE International Conference on Robotics and Automation, Minneapolis, MN, 1996, pp. 884–888.
6. Lujan, J., Parrilla, P., Robles, R. *et al.* (1998) Laparascopic cholecystectomy vs open cholecystectomy in the treatment of acute cholecystitis. *Archives of Surgery*, **133**, 173–175.
7. Fager, P.J. and von Wowern, P. (2004) The use of haptics in medical applications. *International Journal of Medical Robotics*, **1**, 36–42.
8. Gaspari, A. and Di Lorenzo, N. (2003) State of the art of robotics in general surgery. *Business Briefing: Global Surgery*, **1**, 1–6.
9. Sastry, S.S., Cohn, M., and Tendick, F. (1997) Milli-robotics for remote, minimally invasive surgery. *Robotics and Autonomous Systems*, **21**, 305–316.
10. Uchio, Y., Ochi, M., Adachi, N. *et al.* (2002) Arthroscopic assessment of human cartilage stiffness of the femoral condyles and the patella with a new tactile sensor. *Medical Engineering & Physics*, **24**, 431–435.
11. Zamorano, L., Li, Q., Jain, S., and Kaur, G (2004) Robotics in neurosurgery: state of the art and future technological challenges. *International Journal of Medical Robotics*, **1**, 7–22.
12. Cleary, K. and Nguyen, C. (2001) State of the art in surgical robotics: clinical applications and technology challenges. *Computer Aided Surgery*, **6**, 312–328.
13. Rebello, K.J. (2004) Applications of MEMS in surgery. *Proceedings of the IEEE*, **92**, 43–55.
14. Michael, J.M. (2001) Minimally invasive and robotic surgery. *Journal of the American Medical Association*, **285**, 568–572.
15. Dargahi, J. (2002) An endoscopic and robotic tooth-like compliance and roughness tactile sensor. *Journal of Mechanical Engineering Design (ASME)*, **124**, 576–582.
16. Son, J.S., Cutkosky, M.R., and Howe, R.D. (1995) Comparison of contact sensor localization abilities during manipulation. Proceedings of the 1995 IEEE/RSJ International Conference on Intelligent Robots and Systems 95, 'Human Robot Interaction and Cooperative Robots', pp. 96–103.
17. Miyaji, K., Sugiura, S., Inaba, H. *et al.* (2000) Myocardial tactile stiffness during acute reduction of coronary blood flow. *The Annals of Thoracic Surgery*, **69**, 151–155.
18. Dargahi, J., Kahrizi, M., Rao, N.P., and Sokhanvar, S. (2006) Design and microfabrication of a hybrid piezoelectric-capacitive tactile sensor. *Sensor Review*, **26**, 186–192.
19. Dargahi, J. and Najarian, S. (2004) Human tactile perception as a standard for artificial tactile sensing---a review. *International Journal of Medical Robotics*, **1**, 23–35.

20. Dario, P. (1991) Tactile sensing: technology and applications. *Sensors and Actuators A: Physical*, **26**, 251–256.

21. Hertz, D.L.J. and Farinella, A.C. (1998) Shore a durometer and engineering properties. Presented at the Fall Technical Meeting of The New York Rubber Group, pp. 1–13.

22. Qi, H.J., Joyce, K., and Boyce, M.C. (2003) Durometer hardness and the stress–strain behavior of elastomeric materials. *Rubber Chemistry and Technology*, **76**, 419–435.

23. Omata, S. and Terunuma, Y. (1991) Development of new type tactile sensor for detecting hardness and/or software of an object like the human hand. International Conference on Solid-State Sensors and Actuators, 1991. Digest of Technical Papers, TRANSDUCERS '91, pp. 868–871.

24. Yamamoto, Y. and Kawai, K. (1999) Development of measuring method for softness of epidermis using rotational step response. Proceedings of the 16th IEEE Instrumentation and Measurement Technology Conference, 1999, IMTC/99, pp. 359–364.

25. Bajcsy, R. (1985) Shape from touch, Presented at the International Conference on Advances in Automation and Robotics, JAI Press, London.

26. Dario, P., Domenici, C., Bardelli, R. *et al.* (1983) Piezoelectric polymers: new sensor materials for robotic applications. presented at the 13th International Symposium Industrial Robots, Chicago, IL, 1983.

27. De Rossi, D., Nannini, A., and Domenici, C. (1988) Artificial sensing skin mimicking mechanoelectrical conversion properties of human dermis. *IEEE Transactions on Biomedical Engineering*, **35**, 83–92.

28. Thomas, V., Patil, K., and Radhakrishnan, S. (2004) Three-dimensional stress analysis for the mechanics of plantar ulcers in diabetic neuropathy. *Medical and Biological Engineering and Computing*, **42**, 230–235.

29. Tanaka, M., Furubayashi, M., Tanahashi, Y, and Chonan, S. (2000) Development of an active palpation sensor for detecting prostatic cancer and hypertrophy. *Smart Materials and Structures*, **9**, 878–884.

30. Beebe, D.J., Denton, D.D., Radwin, R.G., and Webster, J.G. (1998) A silicon-based tactile sensor for finger-mounted applications. *IEEE Transactions on Biomedical Engineering*, **45**, 151–159.

31. Dargahi, J., Parameswaran, M., and Payandeh, S. (2000) A micromachined piezoelectric tactile sensor for an endoscopic grasper-theory, fabrication and experiments. *Journal of Microelectromechanical Systems*, **9**, 329–335.

32. Shikida, M., Shimizu, T., Sato, K., and Itoigawa, K. (2003) Active tactile sensor for detecting contact force and hardness of an object. *Sensors and Actuators A: Physical*, **103**, 213–218.

33. Lindahl, O.A., Omata, S., and Angquist, K.A. (1998) A tactile sensor for detection of physical properties of human skin in vivo. *Journal of Medical Engineering & Technology*, **22**, 147–153.

34. Miyaji, K., Furuse, A., Nakajima, J. *et al.* (1997) The stiffness of lymph nodes containing lung carcinoma metastases. *Cancer*, **80**, 1920–1925.

35. Wellman, P.S., Howe, R.D., Dewagan, N. *et al.* (1999) Tactile imaging: a method for documenting breast masses. Proceedings of the First Joint Engineering in Medicine and Biology, 21st Annual Conference and the 1999 Annual Fall Meeting of the Biomedical Engineering Society BMES/EMBS Conference, 1999, p. 1131.

36. Wellman, P.S. and Howe, R.D. (1999) Extracting features from tactile maps. Proceedings of MICCAI, 1999, pp. 1133–1142.

37. Wellman, P.S., Dalton, E.P., Krag, D. *et al.* (2001) Tactile imaging of breast masses: first clinical report. *Archives of Surgery*, **136**, 204–208.

38. Barman, I. and Guha, S.K. (2006) Analysis of a new combined stretch and pressure sensor for internal nodule palpation. *Sensors and Actuators A: Physical*, **125**, 210–216.

39. Hosseini, M., Najarian, S., Motaghinasab, S., and Dargahi, J. (2006) Detection of tumours using a computational tactile sensing approach. *The International Journal of Medical Robotics and Computer Assisted Surgery*, **2**, 333–340.

40. Kattavenos, N., Lawrenson, B., Frank, T.G. *et al.* (2004) Force-sensitive tactile sensor for minimal access surgery. *Minimally Invasive Therapy Allied Technologies*, **13**, 42–46.

41. Kandel, E., Schwartz, J., and Jessell, T. (2000) *Principles of Neural Science*, 4th edn, McGraw-Hill.

42. Johansson, R.S. and Vallbo, A.B. (1976) Skin mechanoreceptors in the human hand: an inference of some population properties, in *Sensory Functions of the Skin in Primates* (ed. Y Zotterman), Pergamon Press, pp. 185–199.

43. Johansson, R.S. and Vallbo, A.B. (1979) Tactile sensitivity in the human hand: relative and absolute densities of four types of mechanoreceptive units in glabrous skin. *Journal of Physiology*, **286**, 283–300.

44. Hayashi, K. and Takahata, M. (2005) Objective evaluation of tactile sensation for tactile communication. *NTT DoCoMo Technical Journal*, **7**, 39–43.

45. Johansson, R.S. (1978) Tactile sensibility in man. A quantitative study of the population of mechanoreceptive units in the glabrous skin area of the hand. Medical Dissertation, Umea University, Umea, Sweden.

46. Lindenblatt, G. and Silny, J. (2006) Evaluation and comparison of 50 Hz current threshold of electrocutaneous sensations using different methods. *Journal of Zhejiang University Science B*, **7**, 933–946.

47. Johansson, R.S. and Lundström, R. (1983) Lokala vibrationer och handens taktila känsel, in Arbetarskyddsfondens sammafattningar (re. ASF rapport, proj. 79/104.), L. 1983 Nr. 571, ed. Stockholm.

48. Johansson, R.S. and Vallbo, Å.B. (1983) Tactile sensory coding in the glabrous skin of the human hand. *Trends in Neurosciences*, **6**, 27–31.

49. Vallbo, A.B. and Johansson, R.S. (1978) Tactile sensory innervation of the glabrous skin of the human hand, in *Active Touch, the Mechanism of Recognition of Objects by Manipulation* (ed. G. Gordon), Pergamon Press Ltd, Oxford, pp. 29–54.

50. Kajimoto, H., Kawakami, N., Maeda, T., and Tachi, S. (2001) Electrocutaneous display as an interface to a virtual tactile world. Proceedings of the IEEE Virtual Reality (VR.01), 2001, pp. 289–290.

51. Edin, B.B. and Vallbo, A.B. (1988) Stretch sensitization of human muscle spindles. *Journal of Physiology*, **400**, 101–111.

52. Edin, B.B. and Vallbo, A.B. (1987) Twitch contraction for identification of human muscle afferents. *Acta Physiologica Scandinavica*, **131**, 129–138.

53. Edin, B.B. (1988) Classification of muscle stretch receptor afferents in humans. Medical Dissertation, Umea University, Umea, Sweden.

54. Edin, B.B. and Abbs, J.H. (1991) Finger movement responses of cutaneous mechanoreceptors in the dorsal skin of the human hand. *Journal of Neurophysiology*, **65**, 657–670.

55. Jungmann, M. and Schlaak, H.F. (2002) Miniaturised electrostatic tactile display with high structural compliance. Proceeding of the Eurohaptics, Edinburgh, UK, 2002.

56. Schmidt, R.F. and Tews, G. (1987) *Physiologie des Menschen*, 23th edn, Springer-Verlag, Berlin.

57. Moy, G. and Singh, U. (2000) Human psychophysics for teletaction system design. *The Electronic Journal of Haptics Research*, **1**. 1–20.

58. Kaczmarek, K.A., Webster, J.G., Bach-y-Rita, P., and Tompkins, W.J. (1991) Electrotactile and vibrotactile displays for sensory substitution systems. *IEEE Transactions on Biomedical Engineering*, **38**, 1–16.

59. Tan, H., Radcliffe, J., Book, N. *et al*. (1994) Human Factors for the Design of Force-Reflecting Haptic Interfaces.

60. Caldwell, D.G., Tsagarakis, N., and Giesler, C. (1999) An integrated tactile/shear feedback array for stimulation of finger mechanoreceptor. Proceedings of the 1999 IEEE International Conference on Robotics and Automation, 1999, pp. 287–292.

61. Kode, V.R.C. (2006) Design and characterization of a novel hybrid actuator using shape memory alloy and D.C motor for minimally invasive surgery applications. MSc Dissertation, Department of Electrical Engineering and Computer Science, Case Western Reserve University.

62. Tendick, F., Sastry, S.S., Fearing, R.S., and Cohn, M. (1998) Applications of micromechatronics in minimally invasive surgery. *IEEE/ASME Transactions on Mechatronics*, **3**, 34–42.

63. Saleh, J.W. (1988) *Laparoscopy*, Saunders, Philadelphia.

64. Consensus Development Panel (1993) NIH Consensus conference. Gallstones and laparoscopic cholecystectomy. *Journal of American Medical Association*, **269**, 1018–1024.

65. Way, L.W., Bhoyrul, S., and Mori, T. (1995) *Fundamentals of Laparoscopic Surgery*, Churchill Livingstone, London.

66. Graber, J.N., Schultz, L.S., Pietrafitta, J.J., and Hickok, D.F. (1993) *Laparoscopic Abdominal Surgery*, McGraw-Hill, San Francisco.

67. Silverstein, F.E. and Tytgat, G.N.J. (1991) *Atlas of Gastrointestinal Endoscopy*, Gower Medical, New York.

68. Semm, K. (1987) *Operative Manual for Endoscopic Abdominal Surgery: Operative Pelviscopy: Operative Laparoscopy*, Year Book Medical, Chicago, London.

69. RTI International (2005) News Release – RTI International Awarded Grant to Improve 3D Surgery Technology, www.rti.org (accessed 2012).

70. Voges, U. (1996) Technology in laparoscopy – what to expect in the future. *Urologe Ausgabe*, **35**, 205–214.

71. Tendick, F., Jennings, R.W., Tharp, G.K., and Stark, L.W. (1993) Sensing and manipulation problems in endoscopic surgery: experiment, analysis, and observation. *Presence*, **2**, 66–81.

72. Cohn, M., Deno, C., and Fuji, J. (1993) Hydraulic Actuator, Robot Containing Same, and Method of Producing Same.

73. Wendlandt, J.M. and Sastry, S.S. (1994) Design and control of a simplified Stewart platform for endoscopy. Proceedings of the 33rd IEEE Conference on Decision and Control, 1994, pp. 357–362.
74. Suzumori, K., Iikura, S., and Tanaka, H. (1991) Development of flexible microactuator and its applications to robotic mechanisms. Proceedings of the 1991 IEEE International Conference on Robotics and Automation, 1991, pp. 1622–1627.
75. Wendlandt, J.M. (1994) *Milli Robotics for Endoscopy*, University of California at Berkeley, Department of EECS.
76. Deno, C., Murray, R., Pister, K., and Sastry, S. (1992) *Finger-like Biomechanical Robots*, University of California at Berkeley, Department of EECS.
77. King, C.H., Culjat, M.O., Franco, M.L. *et al*. (2009) Tactile feedback induces reduced grasping force in robot-assisted surgery. *IEEE Transactions on Haptics*, **2**, 103–110.
78. Shadow Robot Company Ltd (2006) Shadow Dextrous Hand, www.shadowrobot.com (accessed 2012).
79. Moy, G., Wagner, C., and Fearing, R.S. (2000) A compliant tactile display for teletaction. Proceedings of the ICRA '00 IEEE International Conference on Robotics and Automation, 2000, vol. 4, pp. 3409–3415.
80. Howe, R.D. and Matsuoka, Y. (1999) Robotics for surgery. *Annual Review of Biomedical Engineering*, **1**, 211–240.

2

Tactile Sensing Technologies

Tactile sensing technologies are employed wherever interactions between a contact surface and the environment are to be measured and registered. As mentioned in Chapter 1, development in tactile sensing technology is driven by the need to provide for an ever increasing number of new applications and requirements, especially in the robotics and medical fields.

2.1 Introduction

There are many physical principles that can be exploited for the development of tactile sensors, but, because the technologies involved are very diverse, this chapter considers only general aspects. The operation of a tactile sensor is not only a function of the sensor structure and its characteristics, but is also dependent on the material of the object being touched or grasped.

Much research into the design and manufacture of a variety of tactile sensors has been reported. For instance, the development of an 8×8 silicon pressure tactile sensor array with on-chip signal readout circuits (Figure 2.1) has been implemented by Wen *et al*. [1]. The integrated sensor array was fabricated by using a combination of micro-electro-mechanical systems (MEMS) and micro-electronics. The prototype device was characterized by the pressure range of 0– 150 kPa, in which the sensors exhibited a linear response with a mean sensitivity of 30.1 mV kPa^{-1}. Another tactile sensor chip has been developed for measuring the distribution of forces on its surface [2]. The chip has eight force-sensitive areas, called 'taxels,' with a pitch of 240 μm. Surface micro-machining techniques were used to produce small cavities that work as pressure-sensitive capacitors. To enable transduction of normal forces to the sensitive areas, the sensor chip surface was covered with silicone rubber. The radius of the sphere, and the load working on it, can be estimated precisely from the tactile sensor output data.

An integrated three-dimensional tactile sensor with robust MEMS structure and soft contact surface suitable for robotic applications has been developed by Mei *et al*. [3] (Figure 2.2).

The sensor has a maximum force range of 50 N in the vertical direction and ± 10 N in the x and y horizontal directions. The tactile sensor includes 4×8 sensing cells, each

Tactile Sensing and Displays: Haptic Feedback for Minimally Invasive Surgery and Robotics, First Edition.
Saeed Sokhanvar, Javad Dargahi, Siamak Najarian, and Siamak Arbatani.
© 2013 John Wiley & Sons, Ltd. Published 2013 by John Wiley & Sons, Ltd.

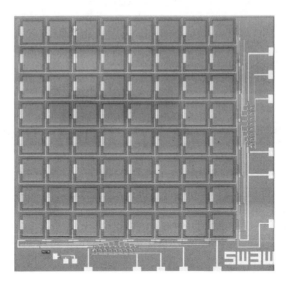

Figure 2.1 Optical micrograph of the fabricated prototype 8 × 8 tactile sensor array [1] (©
Elsevier)

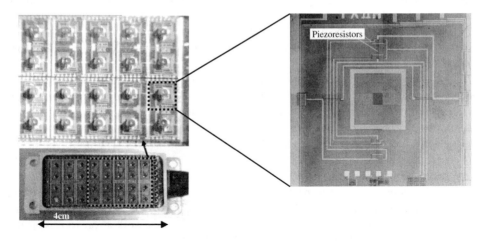

Figure 2.2 Tactile sensor without the rubber surface [3] (© Elsevier)

exhibiting an independent, linear response to the three components of forces applied on the
cells. With neural-network training, the tactile sensor produces reliable three-dimensional
force measurements and repeatable response on tactile images. Another study describes
the development of a new type of tactile sensor which is designed to operate with a
PZT element employed as the sensor [4]. Most researchers have attempted to design
tactile sensors with a number of discrete sensing elements arranged in a matrix form [5].
The main problem with this type of design is crosstalk [6]. Although the force is actually
exerted on a single element of the matrix, any undesirable response from adjacent sensing
elements can cause a measurement error. In cases where polyvinylidene fluoride (PVDF)

film has been used as the basis for the design of a matrix of high spatial resolution tactile sensors, the same crosstalk problems have often been reported [5]. Another problem with a matrix array of PVDF sensing elements is its bulk, due to the fact that a coaxial feed is required for each element, and the potential for instability. These factors combine to make it unsuitable for most robotics applications, so researchers are now concentrating on developing tactile sensors with a minimal number of sensing elements.

In the following sections, various types of tactile sensors will be discussed, together with their Principle of operation and applications.

2.2 Capacitive Sensors

Capacitive sensors make use of the change in capacitance between two electrodes. For example, when pressure is applied, the membrane electrode deflects and changes the gap, and therefore the capacitance, between the two electrodes in direct proportion to the pressure exerted. Dhuler *et al*. [7] designed a silicon-based capacitive pressure sensor, in which the electrodes were made of a planar comb structure. The sensor element comprised two parts: first, a movable elastic structure that transforms a force into a displacement, and second, a transformation unit consisting of electrodes that transform the displacement into a measurable change in capacitance. By measuring the capacitance change on both sides, high linearity and sensitivity were obtained. Compared to piezoresistive sensors, capacitive sensors have better long-term stability, higher sensitivity, and no hysteresis. However, capacitive pressure sensors require more complex signal processing and are also more costly to produce.

2.3 Conductive Elastomer Sensors

Pliable materials that possess defined force–resistance characteristics have received a lot of attention in tactile sensor research. The basic principle of conductive elastomer sensors lies in measuring the resistance of a conductive elastomer (or foam) between two points. The majority of sensors use an elastomer that consists of a carbon-doped rubber. Figure 2.3 shows the schematic of a conductive elastomer sensor.

In the sensor shown in Figure 2.3, the deformation due to an applied force causes a change in density and, subsequently, a change in the resistance of the elastomer.

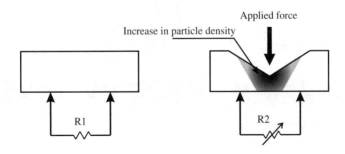

Figure 2.3 Schematic of a conductive elastomer sensor

Conductive elastomer- or foam-based sensors do, however, suffer from some significant disadvantages:

- An elastomer has a long nonlinear time constant. In addition, the time constant of the elastomer when force is applied is different to the time constant when the applied force is removed.
- The force–resistance characteristics of elastomer-based sensors are highly nonlinear, requiring the use of signal processing algorithms.
- Due to the cyclic application of forces experienced by a tactile sensor, the resistive medium within the elastomer migrates over a period of time. Additionally, the elastomer will become permanently deformed and fatigued, leading to permanent deformation of the sensor. In the long-term, its stability will become poor, necessitating replacement after an extended period of use.

Due to the simplicity of design and interfacing of these sensors, even with their electrical and mechanical disadvantages, the majority of industrial analog touch or tactile sensors are, nonetheless, based on the principle of resistive sensing.

2.4 Magnetic-Based Sensors

A magneto-resistive or magneto-elastic material is one whose magnetic characteristics change with an externally applied physical force. Magneto-elastic sensors have a number of advantages, such as high sensitivity and dynamic range, no measurable mechanical hysteresis, a linear response, and physical robustness. Of the two approaches to designing tactile sensors based on magnetic transduction, one is based on the principle that the movement of a small magnet by an applied force will cause the flux density at the point of measurement to change (Figure 2.4). The observed change in flux density can then be measured by using either the Hall effect or a magneto-resistive device. The second approach involves fabricating the core of a transformer or inductor from a magneto-elastic material that will deform under pressure and cause the magnetic coupling between

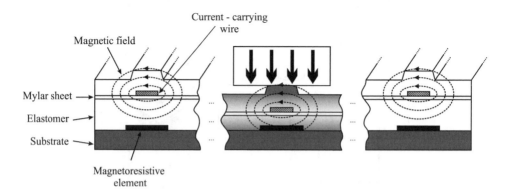

Figure 2.4   Magneto-resistive sensor using current-carrying wires (recreated) [8]

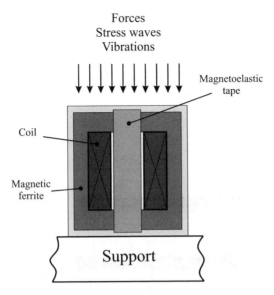

Forces
Stress waves
Vibrations

Magnetoelastic
tape

Coil

Magnetic
ferrite

Support

Figure 2.5 Magneto-elastic sensor. The permeability of a magneto-elastic tape is changed if a force/strain (wave or vibration) is applied to the magneto-elastic tape (recreated) [9]

the transformer windings to change (Figure 2.5). If a very small permanent magnet is held above the detection device by a compliant medium, the change in flux caused by the magnet's movement due to an applied force can be detected and measured. The field intensity follows an inverse relationship, leading to a nonlinear response, but which can be easily made linear by suitable processing. A tactile sensor using magneto-elastic material consists of a material bonded to the substrate and is used as a core for an inductor. As the core is stressed, the susceptibility of the material changes, which is measured as a change in the coil inductance.

2.5 Optical Sensors

The rapid expansion of optical technology in recent years has led to the development of a wide range of tactile sensors. The operating principles of optically based sensors are well known and they fall into two classes: 'intrinsic,' where the optical phase, intensity, or polarization of transmitted light is modulated without interrupting the optical path, and 'extrinsic,' where the physical stimulus interacts with the light external to the primary light path.

Intrinsic and extrinsic optical sensors can be used for touch, torque, and force sensing. For industrial applications, the most suitable will be that which requires the least optical processing. For robotic touch and force-sensing applications, the extrinsic sensor, based on intensity measurement, is the most widely used, due to the simplicity of its construction and subsequent information processing. The potential benefits of using optical sensors are immunity to external electromagnetic interferences, being far from the optical source due to use of optical fibers, and low weight and volume.

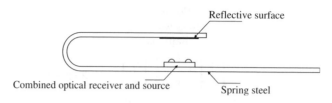

Figure 2.6   Reflective touch sensor

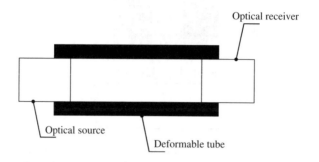

Figure 2.7   Optical sensor shielded by deformable tube

Touch and tactile optical sensors have been developed using a range of optical technologies. The force sensitivity in the optical sensor is determined by a spring or elastomer. For instance, in the schematic sensor shown in Figure 2.6, the distance between the reflector and the plane of the source and the detector is the variable. The intensity of the received light is a function of distance, and hence the applied force. The U-shaped spring is manufactured from spring steel, leading to an overall compact design. This sensor has been successfully used in an anthropomorphic end effector. To prevent crosstalk from external sources, the sensor can be constructed around a deformable tube, resulting in a highly compact sensor. A schematic of an optical sensor covered by a deformable tube is shown in Figure 2.7.

2.6   MEMS-Based Sensors

MEMS is the technology of very small mechanical devices driven by electricity; it merges at the nanoscale into nanoelectromechanical systems (NEMS) and nanotechnology.

MEMS are made up of components between 1 and 100 μm in size (i.e., 0.001–0.1 mm) and MEMS devices generally range in size from 20 μm to a millimeter. They usually consist of a central unit that processes data, the microprocessor, and several components that interact with the outside, such as micro-sensors, to detect temperature, chemical, or magnetic changes, as well as one or more micro-actuators. At these size scales, the standard constructs of classical physics are not always useful. Because of the large surface-area to volume ratio of MEMS, surface effects such as electrostatics and wetting dominate volume effects such as inertia or thermal mass. At the beginning of the 1970s, the producers used engraved substrate plates to produce pressure sensors. At the beginning of the 1980s, research had moved on and the technique of micro-manufacturing was used to create actuators made up of polycrystalline silicon used in disk read heads. At the end of the 1980s, the potential of MEMS was widely recognized and their applications increasingly entered the fields of micro-electronics and biomedicine.

MEMS-based sensors are a class of devices that build very small electrical and mechanical components on a single chip. MEMS-based sensors are a crucial component in automotive electronics, medical equipment, hard disk drives, computer peripherals, wireless devices, and smart portable electronics such as cell phones and PDAs.

For automotive safety, acceleration sensors provide crash detection for efficient deployment of forward and side airbags, as well as other automotive safety devices. Accelerometers are also used in electronic stability control (ESC) to measure the lateral acceleration of the vehicle and to help drivers maintain control during potentially unstable driving conditions. Acceleration sensors are used to detect whether or not the car is moving, in order to save power, and to measure wheel speed and/or direction of rotation. In cell phones, MEMS products activate different features by using the more natural hand movements of tilting, rather than pushing several buttons.

In specialized healthcare monitoring applications, pressure sensors provide key patient diagnostics. MEMS technology utilizes different sensing technologies, including capacitive, piezoresistive, and optical. For example, piezoresistive three-dimensional tactile sensor arrays, suitable for robotic applications, have been developed by Mei *et al*. [3]. A polymer-based MEMS tactile sensor has been used for texture classification by Kim *et al*. [10]. In a novel design by R. Ahmadi *et al*. [11], a magnetic resonance imaging (MRI)-compatible optical-fiber MEMS tactile sensor was developed, without any need to use the array of this sensor to measure the distributed tactile information.

2.7 Piezoresistive Sensors

Change in resistivity of a semiconductor due to applied mechanical stress is called piezoresistivity and is a widely used sensor principle. Nowadays, this principle is used in the MEMS field for a wide variety of sensing applications, including accelerometers, pressure sensors [12], gyro rotation rate sensors [13], tactile sensors [14], flow sensors, sensors for monitoring the structural integrity of mechanical elements [15], and chemical/biological sensors.

In metal, resistance change is caused by an alteration in its geometry (length and cross section) resulting from applied mechanical stress. In semiconductors, in addition to the geometric effect, the resistance of the material itself also changes due to the change in the mechanical state of stress. Consequently, there are two important ways by which the resistance value can change with applied strain. The magnitude of resistance change originating from the piezoresistive effect is much greater than what is achievable from the change in geometry. In particular, doped[1] silicon exhibits remarkable piezoresistive response characteristics among all known piezoresistive materials [16, 17].

The fact that the resistivity of semiconductor silicon may change under applied strain is fascinating. In semiconductors, changes in inter-atomic spacing resulting from strain

---

[1] In semiconductor production, doping is intentionally introducing impurities into an extremely pure (also referred to as intrinsic) semiconductor for the purpose of modulating its electrical properties.

affects the band gap[2], making it easier (or harder, depending on the material and strain) for electrons to be raised into the conduction band. This results in a change in resistivity of the semiconductor.

From the macroscopic point of view, the change in resistance is linearly related to the applied strain, according to:

$$\frac{\Delta R}{R} = G \frac{\Delta L}{L}$$

where R is the resistance, L is the length, and the constant, G, is the gauge factor of a piezoresistor. By rearranging the above equation, an explicit expression for G is derived as follows:

$$G = \frac{\Delta R/R}{\Delta L/L} = \frac{\Delta R}{\varepsilon R}$$

where ε is the strain.

The resistance of a resistor is customarily measured along its longitudinal axis. Externally applied strain, however, may contain three primary vector components, one along the longitudinal axis of a resistor and two arranged 90° to the longitudinal axis and each other, which results in two gauge factors. Longitudinal and transverse gauge factors are different for any given piezoresistive material. For polycrystalline silicon, the transverse gauge factor is generally smaller than the longitudinal one.

The piezoresistive effect of semiconductors has been used for sensor devices employing all kinds of semiconductor materials, such as germanium, polycrystalline silicon, amorphous silicon, and single crystal silicon. Due to the magnitude of the piezoresistive effect in silicon, this semiconductor has attracted much attention in the research and development of sensor devices. Piezoresistors have also been successfully employed in the development of various strain gauges and force sensors. Single-crystal and polycrystalline silicon are used for manufacturing semiconductor strain gauges by selectively doping silicon [16–19].

2.7.1 Conductive Elastomers, Carbon, Felt, and Carbon Fibers

Conductive elastomers are insulating rubbers, either natural or silicone based, which are made conductive by adding particles of conducting or semiconducting materials such as silver or carbon. Most of these forms of conductive rubbers show little change in bulk resistance as they are compressed. However, area of contact and hence inverse contact resistance can be made to vary with applied force. An example of a sensor using conductive elastomers described by Hillis [20] is shown in Figure 2.8. At a certain threshold force, the conductive elastomer makes contact with the electrode. Additional force increases the area of contact and thus reduces the contact resistance.

Larcombe [21] has described piezoresistive sensors constructed by sandwiching carbon felt and carbon fibers between metal electrodes, as shown in Figure 2.9. As the load increases, the carbon fibers are compacted together, making more electrical contacts and reducing the felt resistance. At loads in excess of 5 kg, the area of contact between

---

[2] A band gap or energy gap is an energy range in a solid where no electron states can exist. The band gap is a major factor determining the electrical conductivity of a solid. Substances with large band gaps are generally insulators and those with smaller band gaps are semiconductors.

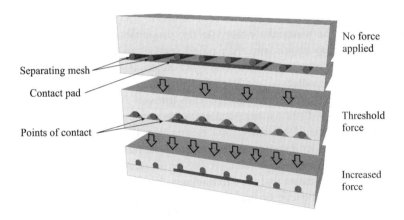

Figure 2.8 Piezoresistive sensor based on conductive elastomer using a separator [8]

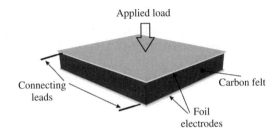

Figure 2.9 Schematic of a carbon felt tactile sensor [8]

touching fibers starts to increase and this leads to a further reduction in resistance. Carbon fiber and carbon felt sensors are rugged and can be shaped. They can also withstand very high temperatures and considerable overloads. One disadvantage of this sensor is that it generates a great deal of electrical noise when the load is less than 10 g. However, these sensors are very robust and well suited for sensing in very inhospitable environments.

2.8 Piezoelectric Sensors

A piezoelectric material is defined as one that either produces an electrical discharge when subjected to a mechanical deformation or undergoes a mechanical deformation when subjected to an electrical input. Polymeric materials that exhibit piezoelectric properties are particularly suitable as tactile sensors.

PVDF is a classic example of a polymer that is now widely being tested for use in tactile sensors. It is used generally in applications that require the highest purity, strength, and resistance to solvents, acids, bases, and heat, and low smoke generation during a fire event. In 1969, the strong piezoelectricity of PVDF was observed by Kawai [22]. The piezoelectric coefficient (d_{33}) of poled thin films of the material were reported to be as large as $6-7$ pC N^{-1} which is 10 times larger than that observed in any other polymers.

PVDF has a glass transition temperature (T_g) of about $-35\ ^\circ$C and is typically 50–60% crystalline. To give the material its piezoelectric properties, it is mechanically stretched to orient the molecular chains and then poled (placed under a strong electric field to induce a net dipole moment) under tension. When poled, PVDF is a ferroelectric polymer, exhibiting efficient piezoelectric and pyroelectric properties. These characteristics make it useful in sensor and battery applications. Thin films of PVDF are used in some newer thermal camera sensors. Unlike other popular piezoelectric materials, such as PZT, PVDF has a negative d_{33} value. Physically, this means that PVDF will compress instead of expand or vice versa, when exposed to the same electric field. The piezoelectric properties of PVDF are used to advantage to manufacture tactile sensor arrays.

PVDF is available in sheets that vary in thickness between 5 μm and 2 mm. It has good mechanical properties and can be molded to an appropriate shape with little difficulty. Metallization is used to apply a thin layer of metal on both sides of the PVDF sheet to form electrodes and collect the charge accumulated. As mentioned earlier, deformation caused by an external force results in an electrical charge that is a function of the applied force. This charge results in a voltage $V = Q/C$, where Q is the charge developed, and C is the capacitance of the device.

Piezoelectric crystals act as transducers that turn force or mechanical stress into electrical charge which, in turn, can be converted into a voltage. A schematic of a piezoelectric sensor is shown in Figure 2.10.

Table 2.1 summarizes and compares different force sensing technologies. Of these technologies, piezoelectricity using PVDF is elaborated in the next chapter. PVDF has shown many outstanding features, such as high sensitivity, wide dynamic response, and the ability to be integrated into MEMS devices, among many others.

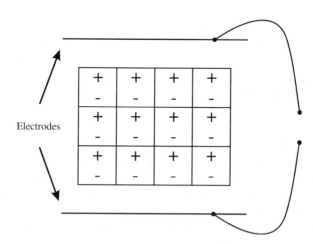

Figure 2.10 Schematic for a piezoelectric sensor

Table 2.1 Comparison of different sensor technologies used for measuring force

Sensor type	Advantages	Disadvantages
Capacitive	Good sensitivity Moderate hysteresis Wide dynamic range Linear response Robust	High impedance Complex circuitry Susceptible to noise Limited spatial resolution Some dielectrics are temperature sensitive
Conductive polymers	Inexpensive Easy to install	High hysteresis Low accuracy Not viable for small force measurements Nonlinear variation
Inductive	High resolution Static and dynamic response	Low frequency response External magnetic field influence
Magnetic (Hall effect)	No temperature dependence Wide dynamic range Low hysteresis Linear response Robust	Miniaturization impossible Measures field in only one direction
Magnetic (magneto-elastic)	Simpler than hall effect Measure field in two directions Wide dynamic range Low hysteresis Linear response Robust	Susceptibility to stray fields and noise

Optical (frustrated internal
   reflection)
	Formable
	Very high resolution tactile
	   image
		Bulky
		Complex construction

Optical (opto-mechanical)	Good repeatability	Creep Memory Hysteresis Temperature dependence
Optical (fiber-optic)	Lower noise Flexible	Complex construction
Optical sensors (photoelasticity)	Linear response Hysteresis Creep Memory	Complicated optic system Not formable

Optical (tracking of optical
   markers)
	No interconnects to break	Requires processing for
		   computing applied force
		Hard to customize

(continued)

Table 2.1 (*continued*)

Sensor type	Advantages	Disadvantages
Strain gauge (metal)	More robust than semiconductor strain gauges	Temperature dependence Small k-factor compared to semiconductor strain gauges
Strain gauge (semiconductor)	Very linear response Low hysteresis Low creep Large k-factor	Vulnerable to overload Not formable Temperature dependence
Piezoresistive (conductive elastomers)	Formable Good gripping surface	Creep Memory Hysteresis Temperature dependence
Piezoresistive(carbon felt and carbon fibers)	Formable Withstand very high temperatures Withstand high overloads	Noise at low loads Not suitable for miniaturizing
Piezoelectric	Wide dynamic range Durability Good mechanical material properties	Fragility of electrical connections Inherently dynamic Difficulty of separating pyroelectric/piezoelectric effects
Pyroelectric	Wide dynamic range Durability Good mechanical material properties	Inherently dynamic Difficulty of separating pyroelectric/piezoelectric effects

References

1. Wen, Z., Wu, Y., Zhang, Z. *et al.* (2003) Development of an integrated vacuum microelectronic tactile sensor array. *Sensors and Actuators A: Physical*, **103**, 301–306.
2. Leineweber, M., Pelz, G., Schmidt, M. *et al.* (2000) New tactile sensor chip with silicone rubber cover. *Sensors and Actuators A: Physical*, **84**, 236–245.
3. Mei, T., Li, W.J., Ge, Y. *et al.* (2000) An integrated MEMS three-dimensional tactile sensor with large force range. *Sensors and Actuators A: Physical*, **80**, 155–162.
4. Omata, S. and Terunuma, Y. (1992) New tactile sensor like the human hand and its applications. *Sensors and Actuators A: Physical*, **35**, 9–15.
5. Lee, M.H. and Nicholls, H.R. (1999) Tactile sensing for mechatronics---a state of the art survey: a review article. *Mechatronics*, **9**, 1–31.
6. Kolesar, E.S., Reston, R.R., Ford, D.G., and Fitch, R.C. (1992) Multiplexed piezoelectric polymer tactile sensor. *Journal of Robotic Systems*, **9**, 37–63.
7. Dhuler, V.R., Mehregany, M., Phillips, S.M., and Lang, J.H. (1992) A comparative study of bearing designs and operational environments for harmonic side-drive micromotors. Micro Electro Mechanical Systems,

1992, MEMS '92, Proceedings. An Investigation of Micro Structures, Sensors, Actuators, Machines and Robot. IEEE, pp. 171–176.

8. Russell, R.A. (1990) *Robot Tactile Sensing*, Prentice Hall, New York, Sydney.

9. ChenYang (2010) Technologies GmbH & Co. KG Magnetoelastic Sensors, www.chenyang-ism.com (2012).

10. Kim, S.H., Engel, J., Liu, C., and Jones, D.L. (2005) Texture classification using a polymer-based MEMS tactile sensor. *Journal of Micromechanics and Microengineering*, **15**, 912–920.

11. Ahmadi, R., Packirisamy, M., Dargahi, J., and Cecere, R. (2012) Discretely loaded beam-type optical fiber tactile sensor for tissue manipulation and palpation in minimally invasive robotic surgery. *IEEE Sensors Journal*, **12**, 22–32.

12. Sugiyama, S., Takigawa, M., and Igarashi, I. (1983) Integrated piezoresistive pressure sensor with both voltage and frequency output. *Sensors and Actuators*, **4**, 113–120.

13. Gretillat, F., Gretillat, M.A., and de Rooij, N.F. (1999) Improved design of a silicon micromachined gyroscope with piezoresistive detection and electromagnetic excitation. *Journal of Microelectromechanical Systems*, **8**, 243–250.

14. Kane, B.J., Cutkosky, M.R., and Kovacs, G.T.A. (2000) A traction stress sensor array for use in high-resolution robotic tactile imaging. *Journal of Microelectromechanical Systems*, **9**, 425–434.

15. Hautamaki, C., Zurn, S., Mantell, S.C., and Polla, D.L. (1999) Experimental evaluation of MEMS strain sensors embedded in composites. *Journal of Microelectromechanical Systems*, **8**, 272–279.

16. Smith, C.S. (1954) Piezoresistance effect in germanium and silicon. *Physical Review*, **94**, 42–49.

17. Yozo, K. (1991) Piezoresistance effect of silicon. *Sensors and Actuators A: Physical*, **28**, 83–91.

18. Toriyama, T. and Sugiyama, S. (2002) Analysis of piezoresistance in p-type silicon for mechanical sensors. *Journal of Microelectromechanical Systems*, **11**, 598–604.

19. French, P.J. and Evans, A.G.R. (1989) Piezoresistance in polysilicon and its applications to strain gauges. *Solid State Electronics*, **32**, 1–10.

20. Hillis, W.D. (1981) *Active Touch Sensing*, Artificial Intelligence Laboratory, Massachusetts Institute of Technology, Cambridge, Massachusetts.

21. Larcombe, M.H.E. (1981) Carbon fibre tactile sensors. Proceedings of the First International Conference on Robot Vision and Sensory Controls, Bedford, pp. 273–276.

22. Kawai, H. (1969) The Piezoelectricity of Poly (vinylidene Fluoride). *Japanese Journal of Applied Physics*, **8**, 975–976.

3

Piezoelectric Polymers: PVDF Fundamentals

In this section, equations are discussed that govern crystals in general, and polyvinylidene fluoride (PVDF) in particular, since it is this medium that is primarily used in sensing applications throught this book.

3.1 Constitutive Equations of Crystals

The constitutive equations for a crystal encompass its mechanical, electrical, and thermal properties and the relationships between these three are illustrated in Figures 3.1 and 3.2 [1]. In the three outer corners, stress, electric field, and temperature are normally chosen as independent variables and all can be thought of as 'forces' applied to the crystal. Alternatively, in the three corresponding inner corners, apparent entropy per unit volume S, electric displacement, and strain, which are the direct results of the 'forces,' can be considered as dependent variables.

The relationships between these pairs of corners (shown by thick lines) are sometimes called principal effects.

The symbols corresponding to each variable and properties are illustrated in Figure 3.2 [1].

1. An increase of temperature produces a change of entropy; thus considering a unit volume:

$$dS = (C/T)\,dT \qquad (3.1)$$

   where C (a scalar) is the heat capacity per unit volume, and T is the absolute temperature.
2. A small change in electric field produces a change in electric displacement according to the equation:

$$dD_i = k_{ij}\,dE_j \qquad (3.2)$$

   where k_{ij} is the permittivity tensor.

Tactile Sensing and Displays: Haptic Feedback for Minimally Invasive Surgery and Robotics, First Edition.
Saeed Sokhanvar, Javad Dargahi, Siamak Najarian, and Siamak Arbatani.
© 2013 John Wiley & Sons, Ltd. Published 2013 by John Wiley & Sons, Ltd.

3. A small change in stress produces a change in strain according to the equation:

$$d\varepsilon_{ij} = s_{ijkl} d\sigma_{kl} \qquad (3.3)$$

where s_{ijkl} are the elastic compliances.

Figures 3.1 and 3.2 also illustrate relations that are called *coupled effects*. These are denoted by the lines joining pairs of points which are not both at the same corner.

In the two diagonal lines at the bottom of the diagram, one shows *thermal expansion*, the strain produced by a change in temperature, while the other shows the *piezocaloric effect*, that is, the entropy (heat) produced by a stress. These coupled effects connect scalars with second-rank tensors and are, therefore, themselves specified by second-rank tensors. As an example, for thermal expansion, the relation is:

$$d\varepsilon_{ij} = \alpha_{ij} dT \qquad (3.4)$$

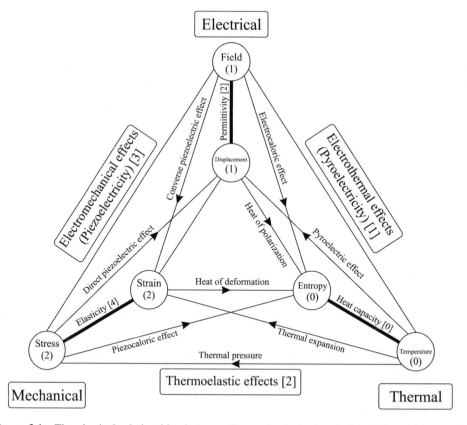

Figure 3.1 The physical relationships between the mechanical, electrical, and thermal properties of a crystal. The tensor rank of each variable is shown in round brackets and the rank of properties in square brackets (By permission of Oxford University Press)

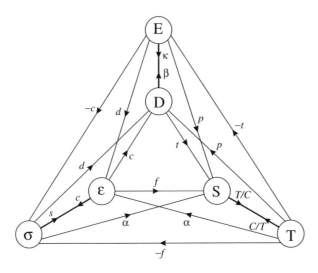

Figure 3.2 The mathematical relationships between the mechanical, electrical, and thermal properties of a crystal (By permission of Oxford University Press)

The piezoelectric coupled effects are shown on the left of the diagram. The direct piezoelectric effect is given in differential form by:

$$dP_i = d_{ijk}d\sigma_{jk} \tag{3.5}$$

since, $D_i = k_0 E_i + P_i$, then $dP_i = dD_i - k_0 dE_i$. Therefore, if the electric field in the crystal is held constant, $dP_i = dD_i$ we may write Equation 3.5 as:

$$dD_i = d_{ijk}d\sigma_{jk}\,(E \text{ constant}) \tag{3.6}$$

The converse piezoelectric effect is the relationship between the electric field and the strain (see Figures 3.1 and 3.2). Since these coupled fields relate to a first-rank tensor (E_i or D_i) and to a second-rank tensor (σ_{ij} or ε_{ij}), they are themselves given by third-rank tensors. The coupled effects on the right of the diagram are concerned with *pyroelectricity*. They all connect a vector (E_i or D_i) to a scalar (S or T), and therefore are expressed by first-rank tensors. The equation for the pyroelectric effect may be written as:

$$dP_i = p_i dT \tag{3.7}$$

where p_i is the pyroelectric coefficient of the crystal and, assuming a constant electric field, we have:

$$dD_i = p_i dT\,(E \text{ constant}) \tag{3.8}$$

The pyroelectricity effect is sometimes categorized into *primary* and *secondary pyroelectricity*. If during the heating a crystal, its shape and size are held fixed (crystal clamped), it is called the primary effect. On the other hand, the crystal may be released so that thermal expansion can occur quite freely. In this case, an extra effect can be observed that is referred to as secondary pyroelectricity and the magnitude of the effects observed in the two experiments would be quite different. Indeed, what is observed in the latter condition is the primary effect plus the additional component of the secondary effect. This phenomenon is not only limited to pyroelectricity, so a similar discussion could equally be given for secondary thermal expansion, or secondary piezoelectricity, if the magnitude of these effects is warranted. Normally the set of σ (stress), E (electric field), and T (temperature) are considered as independent variables, while ε (strain), D (electric displacement), and S (entropy) are dependent variables. However, other alternatives are also possible. Therefore, the general form of linear (first-order effects) constitutive equations for a piezoelectric crystal are given in the following form:

$$\varepsilon_{ij} = s_{ijkl}^{E,T}\,\sigma_{kl} + d_{kij}^{T} E_k + \alpha_{ij}^{E} \Delta T \tag{3.9}$$

$$D_i = d_{ijk}^{T}\sigma_{jk} + k_{ij}^{\sigma,T} E_j + p_i^{\sigma} \Delta T \tag{3.10}$$

$$\Delta S = \alpha_{ij}^{E}\sigma_{ij} + p_i^{\sigma} E_i + \left(C^{\sigma,E}/T\right) \Delta T \tag{3.11}$$

where:

ε_{ij} = second-rank strain tensor

s_{ijkl} = fourth-rank elastic compliance tensor

σ_{ij} = second-rank stress tensor

d_{ijk} = third-rank piezoelectric coefficient tensor (direct and converse effects)

E_k = first-rank electric field tensor

α = second-rank thermal coefficients tensor (thermal expansion and piezocaloric effects)

ΔT = zero-rank temperature tensor

D_i = first-rank electrical displacement

k_{ij} = second-rank permittivity tensor

p_i = first-rank pyroelectric coefficients tensor (pyroelectricity and electrocaloric effects)

(C/T) = zero-rank heat capacity tensor.

The subscripts i, j, k, and l each have a value between 1 and 3. Superscript E means that the material property, in this case the elastic compliance, is measured in a constant electric field (short-circuit condition), and T represents the material property (dielectric constant) measured under constant stress. In general, there are 3^n independent components for each tensor of rank n. For instance, the piezoelectric coefficient matrix (rank 3) and elastic compliance matrix (rank 4) have 27 and 81 independent components, respectively. However, since d_{ijk} are symmetrical in j and k, s_{ijkl} are symmetrical in i, j, k, and l, respectively, independent components can be reduced. For a given piezoelectric material, the number of independent parameters can be further reduced by using symmetry relations in the material. For instance, the piezoelectric PVDF, which is the focus of this section, is classified under an orthorhombic system, class $mm2$ (or C_{2V}) among 32 defined crystal classes. The number of independent variables which are required for this class is depicted in Figure 3.3.

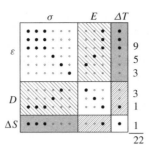

Figure 3.3 The number of independent variables for mm2 class. The darker circles show non-zero components (By permission of Oxford University Press)

To avoid unnecessary complexity when dealing with temperature change and its effects, experiments are normally conducted in an environment in which the temperature is constant. In this case, the third terms of Equations 3.9–3.11 are 0. In many applications, the changes in entropy are so negligible that they can be ignored. For instance, in the majority of sensing and actuating problems, only Equations 3.12 and 3.13 are applicable:

$$\varepsilon_{ij} = s_{ijkl}^{E,T} \sigma_{kl} + d_{kij}^{T} E_k \tag{3.12}$$

$$D_i = d_{ijk}^{T} \sigma_{jk} + k_{ij}^{\sigma,T} E_j \tag{3.13}$$

For sensing applications in the absence of an electric field, these relationships can be simplified even further:

$$\varepsilon_{ij} = s_{ijkl}^{E,T} \sigma_{kl} \tag{3.14}$$

$$D_i = d_{ijk}^{T} \sigma_{jk} \tag{3.15}$$

For uniaxially drawn PVDF, the d_{ij}, s_{ij}, and ε matrices have the following forms [2]:

$$[s] = \begin{bmatrix} s_{11} & s_{12} & s_{13} & 0 & 0 & 0 \\ s_{12} & s_{22} & s_{23} & 0 & 0 & 0 \\ s_{13} & s_{23} & s_{33} & 0 & 0 & 0 \\ 0 & 0 & 0 & s_{44} & 0 & 0 \\ 0 & 0 & 0 & 0 & s_{55} & 0 \\ 0 & 0 & 0 & 0 & 0 & s_{66} \end{bmatrix} \tag{3.16}$$

$$[d] = \begin{bmatrix} 0 & 0 & 0 & 0 & d_{15} & 0 \\ 0 & 0 & 0 & d_{24} & 0 & 0 \\ d_{31} & d_{32} & d_{33} & 0 & 0 & 0 \end{bmatrix} \tag{3.17}$$

$$[\varepsilon] = \begin{bmatrix} \varepsilon_{11} & 0 & 0 \\ 0 & \varepsilon_{22} & 0 \\ 0 & 0 & \varepsilon_{33} \end{bmatrix} \tag{3.18}$$

Because charge is provided by the integration of the electrical displacement over the surface of a body, for a piezoelectric film (including PVDF), the only practical way in which charge, Q, can be collected is on its surface (thickness direction). Therefore, for PVDF films in sensing applications (given a negligible electrical field), we are able to expand Equation 3.15 and combine it with Equation 3.17 to arrive at:

$$Q_3 = d_{31}\sigma_1 + d_{32}\sigma_2 + d_{33}\sigma_3 \qquad (3.19)$$

3.2 IEEE Notation

Another popular alternative form of Equations 3.9 and 3.10 (when the change in temperature ≈ 0) has been suggested by IEEE and adopted by ANSYS:

$$T_{ij} = c_{ijkl}^E S_{kl} - e_{kij} E_k \qquad (3.20)$$

$$D_i = e_{ikl} S_{kl} + \varepsilon_{ij}^S E_k \qquad (3.21)$$

where:

T_{ij} = stress components, (σ_{ij} is adopted in this document)

c_{ijkl} = strain components

e_{kij} = piezoelectric coefficients

E_k = electric led components

D_i = electric displacement components

ε_{ij} = permittivity components.

Implementing a compressed matrix notation, the above equation can be expressed in a more familiar form by replacing ij or kl by p or q, where i, j, k, and l take the values 1, 2, 3 and p, q take the values 1, 2, ..., 6 according to Table 3.1. Then c_{ijkl}, e_{ikl}, and T_{ij} can be replaced by c_{pq}, e_{ip}, and T_p, respectively.

The constitutive Equations 3.20 and 3.21 can be written as:

$$T_p = c_{pq}^E S_q - e_{kp} E_k \qquad (3.22)$$

$$D_i = e_{iq} S_q + \varepsilon_{ik}^S E_k \qquad (3.23)$$

where $S_{ij} = S_p$ when $i = j$, $p = 1, 2, 3$, and $2S_{ij} = S_p$ when $i \neq j$, $p = 4, 5, 6$.

Table 3.1 Conversion table for replacing tensor indices with matrix indices

ij or kl	p or q
11	1
22	2
33	3
23 or 32	4
31 or 13	5
12 or 21	6

In the matrix format they can be written [3]:

$$\{T\}_{6\times 1} = [c]_{6\times 6}\{S\}_{6\times 1} - [e]_{6\times 3}\{E\}_{3\times 1} \tag{3.24}$$

$$\{D\}_{6\times 1} = [e]^T_{6\times 6}\{S\}_{6\times 1} + [\varepsilon]_{6\times 3}\{E\}_{3\times 1} \tag{3.25}$$

Normally the units are: $[e]:\mathrm{pC\ m^{-2}}; [c]:\mathrm{N\ m^{-2}}$ and $[d]:\mathrm{pC\ N^{-1}}$[1]

A typical set of data for the piezoelectric analysis comprises $[e]$, $[c]$, and $[\varepsilon^s]$ or the set of $[d]$, $[s]$, and $[\varepsilon^T]$, where $[\varepsilon^s]$ is the dielectric permittivity matrix at constant strain while $[\varepsilon^T]$ is the dielectric permittivity matrix at constant stress. Between these, the following relationships can be found:

$$\left[\varepsilon^S\right] = \left[\varepsilon^T\right] - [e]^T[d] \tag{3.26}$$

The orthotropic dielectric matrix $[\varepsilon]$ uses electrical permittivity and the following values for the dielectric permittivity matrix at constant stress are used in this book [4, 5]:

$$\left[\varepsilon^T\right] = \begin{bmatrix} 7.35 & 0 & 0 \\ 0 & 9.27 & 0 \\ 0 & 0 & 8.05 \end{bmatrix}$$

The following values are used for the piezoelectric strain matrix [4, 5]:

$$[d] = \begin{bmatrix} 0 & 0 & 0 & 0 & d_{15} & 0 \\ 0 & 0 & 0 & d_{24} & 0 & 0 \\ 20 & 2 & -18 & 0 & 0 & 0 \end{bmatrix} \mathrm{pC\ N^{-1}}$$

and for the stiffness matrix:

$$[d] = \begin{bmatrix} 4.7 & 2.92 & 2.14 & 0 & 0 & 0 \\ 2.43 & 4.83 & 1.99 & 0 & 0 & 0 \\ 2.20 & 2.38 & 4.60 & 0 & 0 & 0 \\ 0 & 0 & 0 & 0.106 & 0 & 0 \\ 0 & 0 & 0 & 0 & 0.104 & 0 \\ 0 & 0 & 0 & 0 & 0 & 2.66 \end{bmatrix} \mathrm{Gpa}$$

In presenting research results for two-dimensional analyses throughout this book, (unless otherwise stated) data was provided by Goodfellow, the manufacturer of the uniaxial and biaxial PVDF film.

3.3 Fundamentals of PVDF

The piezoelectric polymer, PVDF film, exhibits an extremely large piezoelectric and pyroelectric response, which makes it ideally suitable for the design of highly sensitive sensors used in both the robotics and endoscopic environments [6–10]. The piezoelectric applications of the PVDF film vary from robotic (e.g., matrix sensors,

---

[1] Recall from section 3.1 that $D = d\sigma$ and $\varepsilon = d^T E$, where d is the piezoelectric coefficient, σ is the stress, D the electric displacement and E electric field. On the other hand the stress strain is related by: $\sigma = c\varepsilon$ or alternatively $\varepsilon = s\sigma$ where c and s are the elastic stiffness and compliance matrices, respectively. The matrix d is called *piezoelectric strain matrix* as it initially related mechanical strain to the electric field. In addition, from $\varepsilon = d^T E$ we can write $c\varepsilon = cd^T E$ and then using the Hooks relation: $\sigma = eE$. Therefore, the piezoelectric matrix can also be defined in $[e]$ from (*piezoelectric stress matrix*, since it relates the electrical field with stress) where the relationship between them can be defined as: $[e][c][d]$.

displacement measuring transducers) and medical instruments (e.g., blood flow detectors and ultrasonic echography) to military applications (e.g., hydrophones and IR detectors). One of the many reasons why PVDF has proved to be so popular with researchers in the field of tactile sensors is because it is extremely light and inexpensive, and has many wide-ranging applications. Furthermore, it is skin-like, can be prepared between 6 μm and 2 mm in thickness and, because of this flexibility, can be formed into complex surface structures [11]. In addition, PVDF has a bandwidth ranging between 0 Hz and the MHz range and it is primarily due to this that the authors of this book have opted to utilize this medium as the transducer (sensing element) in much of their research and designs.

Although several investigators have reported measurements of the piezoelectric and pyroelectric properties of PVDF film [5, 12–18], in order to consolidate previous findings, some in-depth theoretical and experimental knowledge about PVDF characteristics is still required. Tests are still being carried out on the commercial film supplied by Goodfellow [19] in order to serve two purposes: firstly, to confirm the authenticity of all other PVDF test procedures against those values provided by the manufacturer and, secondly, to provide a base measurement for any future sensor system using PVDF film.

3.4 Mechanical Characterization of Piezoelectric Polyvinylidene Fluoride Films: Uniaxial and Biaxial

Uniaxial and biaxial PVDF films have three and two principal directions, respectively. A traditional tension test machine can be used to measure the mechanical properties (Young's modulus) of the films. Since the uniaxial film is mechanically drawn in one direction, its mechanical properties in that direction are expected to be different than that of the transverse direction. For the biaxial film, however, similar mechanical properties for in-plane directions are expected. The following is a brief description of the machine and method used for these measurements.

A precise mechanical testing machine (Bose, 3200 Series) was used to obtain the stress–strain characteristics of the uniaxial PVDF film on two axes. The peak force of the system was ±225 N in a 12.5 mm range of displacement. The apparatus is shown in Figure 3.4. The extension of the film was measured using a calibrated linear variable differential transformer (LVDT). Three specimens of 110 μm films were cut into a dumbbell shape. The gauge length of each specimen was 26 mm with a width of 8 mm.

Tensile stress–strain characteristics were obtained for specimens oriented in two directions (see below). These averaged results are shown in Figure 3.5. The load (displacement) was applied in a ramp fashion (0.1 mm s^{-1}), and the extension and load were measured during the test. These measurements showed that the Young's modulus, parallel to the drawn direction, was 2.3 GPa, while in the perpendicular to the drawn direction it was 1.85 GPa. The measured modulus values compared favorably with the modulus ranges given by Goodfellow, which were, respectively, 1.8–2.7 GPa for the longitudinal and 1.7–2.7 GPa for the transverse directions. Similar experiments were conducted to measure the mechanical properties of biaxial film (Youngs modulus), for which the result obtained, 1.95 GPa, compared favorably with the value of 2 GPa provided by the manufacturer.

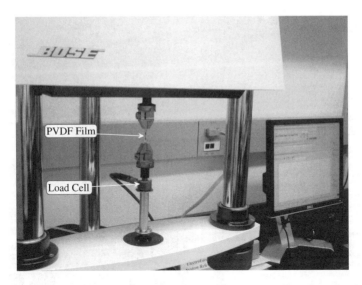

Figure 3.4 The material testing machine used for the characterization of mechanical properties of PVDF films

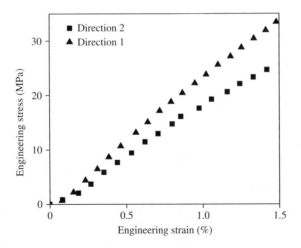

Figure 3.5 Tested stress–strain characteristics for the uniaxial piezoelectric PVDF in direction 1 as well as direction 2

3.4.1 The Piezoelectric Properties of Uniaxial and Biaxial PVDF Films

The three piezoelectric coefficients, d_{31}, d_{32}, and d_{33} are of special interest in sensor applications. The measurement of piezoelectric coefficients, d_{31} and d_{32} compared to d_{33} is straightforward and is explained below. Measurement of d_{33} is more complex and is discussed separately in the next section.

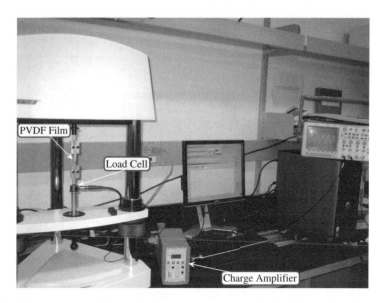

Figure 3.6 The experimental setup used for measurement of piezoelectric coefficients d_{31} and d_{32}

3.4.1.1 Measurement of d_{31} and d_{32}

Since there is no electrical (charge) output for piezoelectric materials in response to applied DC loads, the conventional method of measuring piezoelectric coefficients is to apply periodic loads. A sinusoidal force is the simplest waveform for this purpose and the frequency of this load should, preferably, be higher than its cut-off frequency.[2] The experimental setup is illustrated in Figure 3.6.

In order to measure d_{31}, samples of the piezoelectric coefficient of (Goodfellow) PVDF film were incised in such a way that the drawn direction of the film was parallel to their length. Alternatively, in order to measure d_{32}, samples were cut from the film in such a way that their drawn directions were perpendicular to the length of the samples.

The effective length and width of the 110 μm thick sample were, respectively, 20 and 8 mm. In order to calculate coefficients d_{31} and d_{32}, as shown in Figure 3.7, both the stress and charge density were required. The tensile stress was calculated from the applied force and cross-sectional area of the film. By knowing the amount of charge and also the surface area of the film (metalized area), charge density was calculated. The coefficient d_{31} is the ratio of charge density to the tensile stress.

These measurements conformed favorably with the nominal values quoted by Goodfellow [19], with $d_{31} = 19.0 \pm 0.3$ pC N^{-1} and $d_{32} = 2.0 \pm 0.1$ pC N^{-1}. (The values quoted by Goodfellow were $18 - 20$ pC N^{-1} and 2 pC N^{-1} for d_{31} and d_{32}, respectively.)

---

[2] Piezoelectric PVDF can simply be modeled as a voltage source in series with a capacitor (C). Alternatively, the input impedance of an amplifier can be modeled as being a resistor (R) in which the combination of both R and C acts as a high-pass filter. At a specific frequency (the cut-off frequency), this resistance R and capacitance C causes an attenuation of 0.707 compared to the input amplitude so, in order to avoid this, the normal working frequency must always be greater.

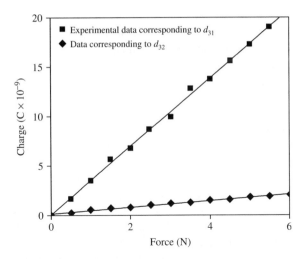

Figure 3.7 The results obtained for the piezoelectric coefficients d_{31} and d_{32} of uniaxial piezoelectric PVDF in direction 1 as well as direction 2, respectively

3.5 The Anisotropic Property of Uniaxial PVDF Film and Its Influence on Sensor Applications

As previously mentioned, the mechanical process on uniaxial and biaxial PVDF films differ insofar as the behavior of uniaxial PVDF film is quite similar to orthotropic materials, whereas biaxial PVDF film exhibits the properties of transversely isotropic materials. In addition, there is no standard process for preparation of PVDF film that causes parity in its mechanical and electrical properties. One of the obstacles encountered during the studies of the anisotropic properties of the piezoelectric PVDF, was the discrepancy observed between the few reported values of orthotropic PVDF due to the fact that it has its roots in dissimilar mechanical and electrical production processes. Various electrical and mechanical properties for uniaxial and biaxial PVDF films are reported by different manufacturers [19–21] and are given in Table 3.2. In this table, piezoelectric and elastic coefficients are given in pC N^{-1} and GPa, respectively. The symbol d_{3h}, represents the hydrostatic piezoelectric coefficient.

This section elaborates on the effects of anisotropic behavior of uniaxial PVDF on the output voltage of uniaxial PVDF and comparing it with the performance of biaxial PVDF in the sensory mode. The behavior of PVDF films can be attributed to their

Table 3.2 PVDF properties reported by manufacturers

| | Uniaxial PVDF | | | | | | | Biaxial PVDF | | |
	d_{31}	d_{32}	d_{33}	d_{3h}	E_{11}	E_{22}	E_{33}	$d_{31} = d_{32}$	d_{33}	$E_{11} = E_{22}$
Piezoflex	14	2	−34	−18	2.5	2.1	0.9	–	–	–
Goodfellow	18/20	2	−20	−6	1.8/2.7	1.7/2.7	–	8	15/16	2
Piezotech	18	3	−20	–	–	–	–	7	−24	–

mechanical anisotropic characteristic, which affects the stress–strain relation, as well as the non-equality, of d_{31} and d_{32} in electromechanical relations.

The response of the sensation and actuation modes varies as a function of the deviation angle, θ which is the angle between the material coordinate system, in which all material properties are defined, and the global coordinate system. In order to show the differences between two types of PVDF films, a simple cantilever beam was selected as the host structure.

Figure 3.8a shows the case where a PVDF film is adhered to a beam in such a way that the material coordinate system and the global coordinate system have the same orientation, while Figure 3.8b represents the case in which axis 1 of the PVDF is rotated by θ with respect to the global coordinate system (x, y, z). As emphasized in Figure 3.8c, the edges of the piezoelectric patches remain parallel to the global axes, that is, the drawn direction (axis 1) rotates with respect to the global axis. The best sensing and actuation response is achieved from a piezoelectric when axis 1 is aligned with the global x- axis. However, in some cases, the orientation of the applied load might result in a two-dimensional stress profile, making it even more important to find the best orientation in which to attach the piezoelectric film. Rotation of the material relative to the global coordinate system can also occur inadvertently during the manufacturing phase of a sensor. Evidently, by simply considering PVDF as an isotropic material, no alteration in the responses in terms of deviation angle would be observed.

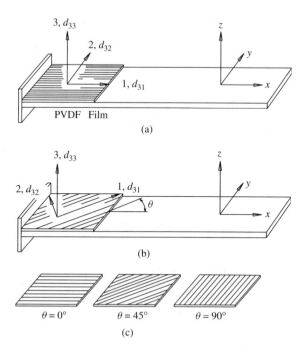

Figure 3.8 The PVDF material coordinate system (1, 2, 3) versus the global coordinate system (x, y, z)

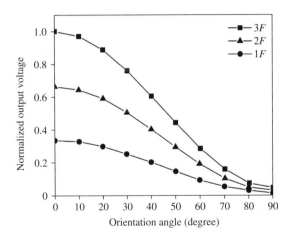

Figure 3.9 The predicted variations of normalized output voltage of the uniaxial piezoelectric PVDF film for different applied loads. For this simulation load F was 10 N

As mentioned earlier, the behavior of uniaxial PVDF film under similar conditions is different to that of biaxial PVDF film. Figure 3.9 shows the predicted variations of normalized total output voltages in terms of the deviation angles of uniaxial PVDF subjected to a set of forces, F, $2F$, and $3F$, in which F is a downward point load of 10 N applied at the tip of the cantilever. The length, width, and thickness of the cantilever are 24 cm, 2.4 cm, and 2 mm, respectively. In Figure 3.9, the curve corresponding to $3F$ is normalized and is used as a reference for other force values. Even though the results are shown for a force $F = 10$N, the trend is identical for any force F in the linear range. The results show that for each force state, the variation of the voltage, due to the orientation angle, is nonlinear and significant (the voltage output at $\theta = 90°$ drops to 4% of its initial value at $0°$).

This significant decrease is due not only to the difference between the piezoelectric coefficients in directions 1 and 2, but also because of the difference in Young's modulus of PVDF in these two principal directions. Despite a nonlinear decrease in output voltage due to the increase in deviation angle, the output voltage increases linearly with increasing applied force for each specific angle. Figure 3.10 shows the contribution of each piezoelectric coefficient in the total response when the deviation angle varies between $0°$ and $90°$. These results are shown in Figure 3.11, in which the variation of output voltage of the uniaxial piezoelectric PVDF film versus force amplitude for different PVDF deviation angles is shown, and the total voltage (V_{total}), as well as its components associated with coefficients $d_{31}(V_{d_{31}})$ and d_{32} ($V_{d_{32}}$), are plotted separately. It can be seen that the total output voltage is mainly comprised of the voltage associated with d_{31}. For example, when $\theta = 0°$, only 1.5% of the total voltage is due to d_{32}, whose reduced contribution to the output voltage can be attributed to the combined effect of mechanical anisotropic properties of the uniaxial PVDF film, as well as the significant difference between the values of d_{31} and d_{32} in the uniaxial PVDF.

The lower Young's modulus in the transverse direction yields a lower stress, hence a lower charge output for the same strain value compared with the biaxial film. It can

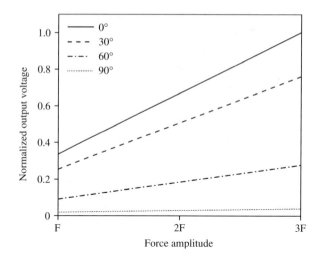

Figure 3.10 Variation of output voltage of the uniaxial piezoelectric PVDF film versus force amplitude for different PVDF deviation angles

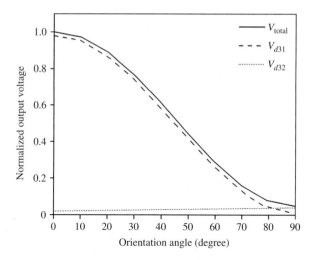

Figure 3.11 Variation of output voltage of the uniaxial piezoelectric PVDF film versus force amplitude for different PVDF deviation angles. The total voltage (V_{total}), as well as its components associated with coefficients d_{31} (V_{d31}) and d_{32} (V_{d32}) are separately plotted

be seen that the maximum difference between the total output voltage and the voltage induced by d_{31} is at $\theta = 90°$ and is equal to 4.5%. The voltage component resulting from the d_{33} coefficient is close to 0 which is similar to biaxial PVDF film behavior. In many applications in which the uniaxial PVDF film experiences merely a uniaxial tension, it is preferable to affix the film in such a way that its material coordinate system coincides with the global coordinate system, where the deviation angle is 0.

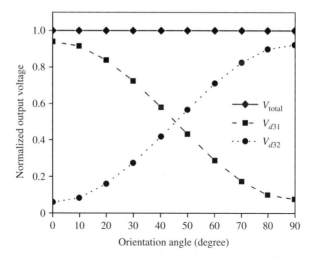

Figure 3.12 Variation of output voltage of the uniaxial piezoelectric PVDF film versus force amplitude for different PVDF deviation angles

3.6 The Anisotropic Property of Biaxial PVDF Film and Its Influence on Sensor Applications

Since the piezoelectric coefficients of biaxial PVDF are equal in the axis 1 and transverse axis 2 directions, this provides a unit ratio of d_{31}/d_{32} for biaxial PVDF film. This equality of piezoelectric coefficients is in compliance with the isotropic mechanical property of biaxial PVDF film (i.e., $E_1 = E_2$). Figure 3.12 shows the result of the output voltage versus the orientation angle of the biaxial PVDF film. The output voltage is normalized with respect to the total output voltage, V_{total}. The variations in the contributing components in the total output voltage, namely $V_{d_{31}}$ and $V_{d_{32}}$ due to the piezoelectric coefficients d_{31} and d_{32} respectively, are clearly shown in this figure. It should be noted that, in order to compute the voltage component associated with each piezoelectric coefficient, the other coefficient were assumed to be 0. From this, it can be interpolated that the total output voltage in biaxial PVDF film is constant and is hence independent of any deviation angle.

The symmetry observed in Figure 3.12 is due to the equality of d_{31} and d_{32} in the two perpendicular directions along with the isotropic material properties of the biaxial PVDF film. It is observed that in the 0° deviation, the contribution of d_{32} to the total output is about 7%. It should be noted that the contribution of d_{33} to total output voltage depends on the boundary conditions and, when there is no constraint on the z-axis, neither is there any stress developed in this direction so there will be no contribution due to d_{33} in the total output.

3.7 Characterization of Sandwiched Piezoelectric PVDF Films

Despite the dominant applications of PVDF film in the extensional mode, there are some situations in which PVDF must be placed between two surfaces. For instance, in traditional piezoelectric force sensors, it is customary to place the piezoelectric sensing element between two plates that also act as electrodes. These plates transmit the normal force to the surface of the PVDF such that they transform the applied point load to a distributed load over the piezoelectric surface. Traditionally, to avoid the complexity of considering friction forces, the sensor package is treated as being a black box in which the relationship between the input and output is considered. Therefore, for a given set of piezoelectric elements and surfaces, the output of the sensor is empirically calibrated in terms of the input load.

However, in this case, the output charge of the PVDF is a combination of the thickness mode charge and another component, caused by the friction force. Therefore, the response of such piezoelectric sensor is highly sensitive to the surface condition and varies with any changes in the manufacturing process that may affect the surface.

As can be seen in this section, the results of this study are helpful in modification and optimization of commercial piezoelectric force sensors. Although this research is performed on piezoelectric PVDF film, the results could equally be applied to any other piezoelectric force sensor in which the friction force cannot be neglected. This study also has another fundamental application in the measurement of d_{33}, namely the piezoelectric coefficient in the thickness mode. This coefficient is difficult to measure because it is extremely difficult to apply a normal force to the film, yet, at the same time, not constrain the lateral movement of the film, which would otherwise induce other stresses that can cause faulty output readings. In order to avoid difficulties associated with the direct measurement of d_{33}, most researchers use two indirect ways of calculating this value [14, 16, 18, 22]. In one method, d_{33} is determined by using the converse piezoelectric effect in which a known electric field is applied and the change in thickness of a small sample is measured. The problem with this approach is in mounting the sample in such a way that its lateral motion is not restricted, which, otherwise, could affect the accuracy of the measurement [18]. In the second method, d_{33} is measured indirectly by measuring the hydrostatic piezoelectric coefficient d_{3h}. Using this value, and knowing the values of d_{31} and d_{32}, the value of d_{33} can be calculated. For a hydrostatic pressure P, the amount of charge is related to all three coefficients (Kepler and Anderson [18]): $\Delta Q/A = -(d_{31} + d_{32} + d_{33})P$, in which $-(d_{31} + d_{32} + d_{33}) = d_{3h}$.

This section introduces a new approach to this problem. The effect of friction on the PVDF output can be characterized by finding the trend of variations for some known friction coefficients; it is then possible to calculate d_{33} in those cases where friction is almost 0. Therefore, a finite element (FE) contact analysis was performed in which piezoelectric PVDF was considered as being an orthotropic material in which the coefficient of friction was varied between 0 and 1 and the output recorded. The results were validated by performing an experiment on a similar geometry using pre-characterized surfaces. It was found that the inverse procedure can also be used in determining the friction coefficient of surfaces. This method confers many advantages over traditional friction measurement methods, such as *in situ* friction measurement, being non-invasive, low weight, and cost.

When a PVDF film is compressed between two rigid flat surfaces, assuming no friction between the surfaces and the PVDF films, the film is free to expand laterally, that is, in the 1 and 2 directions. The output charge can thus be calculated from Equation 3.19 as:

$$Q = d_{33}F_n \tag{3.27}$$

where F_n is the normal applied load. This assumption, however, is difficult to use in practice. Frictional force always exists and causes unwanted components in the output charge. Therefore, in the presence of such a friction-inducing component, the total output charge Q, as defined in Equation 3.27, is different to that of Q_3, as defined in Equation 3.19. Some authors have defined a new symbol, d_{33}^* [21], which relates the applied normal load to the output charge and is not necessarily equal to d_{33}, except in a frictionless state. The contribution of the friction component to the total output depends on several factors, including the magnitude of the applied normal load, the friction coefficient between contact surfaces and the contact area of the PVDF film.

In order to obtain the output of the sandwiched PVDF film in the presence of friction, a contact FE model was developed in ANSYS and similar experiments were also conducted in order to validate the theoretical results of this analysis. In the following sections, details of the modeling, as well as the experimental work, are explained.

3.8   Finite Element Analysis of Sandwiched PVDF

Since FE contact problems in which friction involves, produces non-symmetrical stiffness, it is preferable to use a symmetrization algorithm designed to perform speedy and efficient calculations such as those developed by Laursen and Simo [23] and ANSYS Inc. [24]. One approach to solving contact problems is the Pure Penalty method, which requires matrices showing contacts containing normal and tangential stiffness values. Higher stiffness values decrease the amount of penetration, but can lead to ill-conditioning of the global stiffness matrix and convergence difficulties. The stiffness should be high enough so that contact penetration is acceptably small, but low enough so that the problem will be well-behaved in terms of convergence. The contact traction vector is defined as:

$$\begin{Bmatrix} P \\ \tau_x \\ \tau_y \end{Bmatrix}$$

where:

P = normal contact pressure

τ_x = tangential contact stress in the x-direction

τ_y = tangential contact stress in the y-direction.

The contact pressure is:

$$P = \begin{cases} 0 & | \ u_n > 0 \\ K_n u_n, & | \ u_n \le 0 \end{cases}$$

where K_n and u_n are contact normal stiffness and contact gap size, respectively. The frictional stress is obtained by Coulomb's law:

$$\tau_x = \begin{cases} K_s u_x, & | \tau < 0 \\ \mu K_n u_n, & | \tau = 0 \end{cases}$$

where $\tau = \sqrt{\tau_x^2 + \tau_y^2} - \mu P$

K_s = tangential contact stiffness

u_x = contact slip distance in x direction

μ = frictional coefficient.

It can be seen that the isotropic Coulomb friction model is used in this analysis. In this model, two contacting surfaces can carry shear stresses up to a certain magnitude across their interface before they start sliding relative to each other. This state is known as sticking. The developed shear stress is the function of friction coefficient and contact normal pressure P, so that $\tau = \mu P$. In addition, there could be an initial cohesion sliding resistance b, which is referred to as COHE in ANSYS and can be added to the equivalent shear stress. Therefore, the Coulomb friction model can be defined as $\tau = \mu P + b$. Although contact cohesion provides sliding resistance, even with zero normal pressure, there is, conversely, a maximum contact friction stress, $\tau_{\lim}$, at which, regardless of the magnitude of normal contact pressure, sliding will occur.

The geometry of the modeled problem is shown in Figure 3.13. The computational time for this model can be reduced by taking advantage of the symmetry of the problem and only modeling half of the geometry. For the structural part, that is, the top and bottom plates, the PLANE42 element, was used. This element is defined by four nodes having two DOFs, U_x and U_y, at each node. In order to model the piezoelectric PVDF film, the PLANE223, which is an eight-node coupled-field solid element, was used.

For the piezoelectric analyses, this element has three DOFs at each node, including U_x, U_y, and Volt. The contacts between the PVDF film and the top and bottom surfaces were modeled using TARGE169 which was used to represent various target surfaces associated with the contact elements, CONTA171. The latter element is a 2D, two-node surface-to-surface contact element that represents the deformable surface of a contact pair and is applicable to 2D structural and coupled-field contact analysis. This element is associated with the 2D target segment element (TARGE169) defined with a set of shared real constants. Using the real constants, various controlling parameters such as COHE (b), the cohesion sliding resistance, TAUMAX ($\tau_{\lim}$), and the maximum contact friction

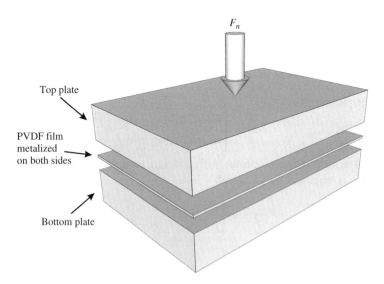

Figure 3.13 The geometry of the problem in which a PVDF film is sandwiched between two solid plates and a load is applied in the normal direction

can be set. The coefficient of friction between the contact surfaces can be introduced as the material property of the contact element, CONTA171.

3.8.1 Uniaxial PVDF Film

The uniaxial PVDF film exhibits orthotropic properties which originate from the mechanical processing it undergoes. This orientation dependency exists in both the mechanical and electrical domains such that the piezoelectric coefficients in directions 1 and 2 are not the only unequal properties, since the Young's moduli of the uniaxial PVDF film in directions 1 and 2 (E_1 and E_2, respectively) also differ.

In this simulation, the PVDF film was a square 30 mm × 30 mm in which the bottom surface was electrically grounded and the total area of the top surface was electrically coupled together to create an equipotential electrode area so, in this case, the whole PVDF film acted as one sensing element. Then a distributed load of 100 kPa was applied through the top plate and the coefficient of friction between each contact surface was increased incrementally, as shown in Figure 3.14. For a better understanding, the negative output voltage is inverted and is shown in the positive y-axis. Although the magnitude of μ can, theoretically, reach very large values, a practical range of $\mu < 1$ was selected for these simulations. It could be seen that the maximum output voltage is obtained at the frictionless state ($\mu = 0$). The output voltage decreases nonlinearly and drops by more than 70% when the coefficient of friction is equal to 0.4. This significant attenuation is associated with the friction-induced voltage component so that for a state in which no friction exists, the PVDF film can freely expand, such that the only nonzero stress component would be σ_3 and the other principal stresses would be 0 ($\sigma_1 = \sigma_2 = 0$). However, when $\mu \neq 0$, the friction force reduces the lateral expansion of the PVDF film, which will develop stress in the 1 and 2-directions. The reason that the inclusion of the x and y components causes a reduction in the total output voltage is due to

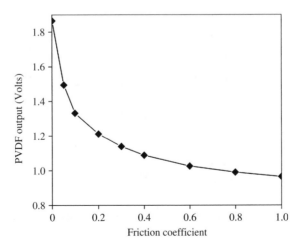

Figure 3.14 The total output voltage of a uniaxial PVDF film compressed between two blocks in the presence of friction

the fact that d_{31} and $d_{32} > 0$ while $d_{33} < 0$. The results of the simulation for $\mu > 1$ show that the variation in the output voltage is negligible. It is important to note that maximum gain can be obtained at low friction for uniaxial PVDF film in the sandwiched configuration. Figure 3.14 also reveals the dependency of maximum output voltage on the friction coefficient, particularly for $\mu < 0.1$. This explains why accurate measurement of d_{33}, using the direct method, is so difficult. On the other hand, this behavior could be beneficial for the accurate measurement of low friction forces and coefficients.

Figure 3.14 illustrates the results for a state in which no friction exists ($\mu = 0$) and conforms to classical theory. Since the metalized piezoelectric PVDF film can be modeled as a capacitor, the relationship between the generated charge and the potential across its electrodes can be written as: $Q = CV$ in which C is the capacitance of the PVDF specimen. Conversely, the capacitance of a PVDF film with area A and thickness t can be calculated from the known relationship: $C = \varepsilon_0 \varepsilon_r A/t$, in which $\varepsilon_0 = 8.854 \times 10^{-12}$ F m^{-1} is the permittivity of free space and ε_r, the relative permittivity of PVDF, is 12. Substitution of the above relations into Equation 3.27 reads:

$$V = \frac{d_{33}t}{\varepsilon_0 \varepsilon_r} P \qquad (3.28)$$

in which $P = F_n/A$ is the applied pressure on the PVDF film. The numerical value of voltage in Equation 2.28 for the data given for the present case is 1.864, which agrees with the result in Figure 3.14 obtained for $\mu = 0$.

In order to elaborate upon the results obtained in our previous analysis, a second simulation was performed in which the top surface of the PVDF film was covered with 10 electrodes of equal width that were evenly positioned on the surface. The bottom surface of the PVDF film was considered as the common ground for all the electrodes. The results of this model under a constant load are depicted in Figure 3.15, which shows

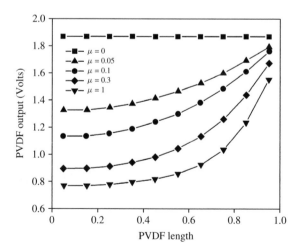

Figure 3.15 The relationship between the output voltage of each segment and different friction coefficients

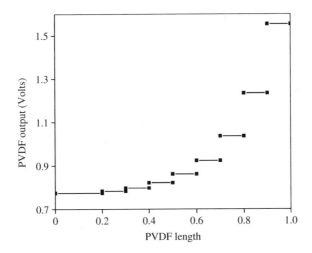

Figure 3.16   Voltage distribution of each sensing elements (for $\mu = 1$)

the output voltages of 10 equal PVDF sensing elements. While the applied load is kept constant, the friction coefficient is varied between 0 and 1. The horizontal axis of the graph is normalized by half the length of the PVDF film. It should be emphasized that each point in this figure represents a constant voltage over the area of a sensing element.

For further clarification, Figure 3.16 shows the distribution of output voltage over the uniaxial PVDF film segments compressed between two plates in the presence of friction and voltage variations from one plate to the other. The horizontal axis of the graph is normalized by half the length of the PVDF film.

From Figure 3.16, it can be seen that the output voltage of each sensing element is a constant value over its area. Since half of the PVDF is modeled, zero is related to the center of the film and one is related to the edge of PVDF. For the frictionless case, that is, $\mu = 0$, the voltage of all the PVDF sensing elements are equal and at their maximum value. However, with an increase in the friction coefficient, the voltage output of each sensing elements begins to reduce. For a given friction coefficient, μ, the voltage attenuation of those elements that are closer to the edges is lower, due to the fact that the outer sensing elements carry less frictional load than those of the middle sensing elements. Therefore, the frictional voltage component of the outer sensing elements is lower, and hence the total output is higher. The trend observed in Figure 3.15 is exactly the same as the trend for the variation of σ_x along the length of the PVDF film at the contact surface. It is now clear that Figure 3.14 represents the average voltage of a charge distribution over the whole PVDF film. On the other hand, the data shown in Figure 3.15, which is the result of segmenting the PVDF into 10 separate sensing elements, are especially helpful to elaborate the study of the distribution of the frictional forces over the contact area.

Having observed the nonlinear behavior of the output voltage of the PVDF against the friction coefficient, the relationship between the amplitude of the applied normal forces and the total output should also be examined. In this simulation, similar to the first test, the

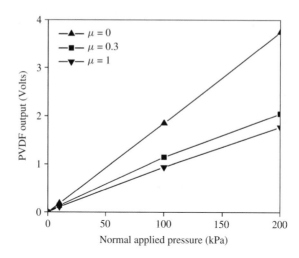

Figure 3.17 The response of PVDF film against applied pressure for different friction coefficients

electrical boundary conditions are defined in such a way that the whole PVDF film forms only one sensing element. Therefore, for each given load, there exists only one voltage output. The applied load ranges from 1 to 200 kPa and the results for three typical friction coefficients, μ: 0, 0.3, 1 are plotted. As demonstrated in Figure 3.17, for a constant friction coefficient, there is a linear relationship between the normal applied load and the sensor output voltage. This justifies why industrial sensors are performing well and linearly, even though the frictional forces have not been taken into consideration. These results also show that the outputs of identical sensors are the same if the friction coefficients of the sensor surfaces are also exactly the same. This could be one of the main reasons for the discrepancy in output voltage of sensors that are similar in structure, but have different contact frictions which can, quite probably, occur during the manufacturing phase. This figure also shows that if, for any reason, uniformity of manufactured sensors cannot be attained, it is better to work in the high-friction region. In this way, the dependency of the sensors on the surface friction will be reduced to a minimum. However, if the contact surface is too rough, then the projections on the surface might cause unexpected results and even damage the PVDF film.

3.8.2 Biaxial PVDF Film

As mentioned earlier, there are some differences between biaxial and uniaxial piezoelectric PVDF films. For instance, biaxial film exhibits isotropic properties in terms of both mechanical ($E_1 = E_2$) and electrical ($d_{31} = d_{32}$) performance. Another important factor is the opposite polarity of d_{33} in uniaxial and biaxial PVDF films. The manufacturer [19], reports the values of d_{33} for the uniaxial and biaxial PVDF films as being -20 and 15 pC N^{-1}, respectively.

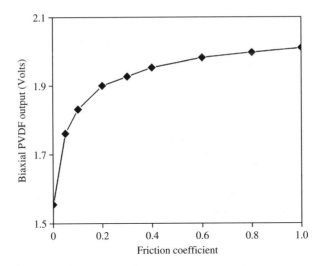

Figure 3.18 The total output voltage of a biaxial PVDF film compressed between two plates in the presence of friction

Since all the piezoelectric coefficients of biaxial PVDF film have the same polarity, compared with the results of the previous simulations, it is expected that the friction-induced components cause an increase in total output. Figure 3.18 demonstrates the relationship between the output voltage of the biaxial piezoelectric PVDF film and the friction coefficient for a constant normal load. It can be seen that this behavior is different to that of uniaxial film. For the biaxial film the friction-induced components cause an increase in total output.

These results could provide some guidance on the proper selection of PVDF film type (between uniaxial and biaxial) for a particular application. The extreme sensitivity of the output voltage in the low friction-coefficient range is again evident. Similar to the analysis of the uniaxial film, the whole length of the biaxial PVDF film is formed into 10 side-by-side discrete electrodes which make 10 independent sensors. As shown in Figure 3.19, voltage of each segment depends on the location of the sensing element, as well as the friction coefficient. In this simulation, the PVDF bottom surface was taken as common ground of all the sensing elements. In accordance with classical theory, and also with the results obtained for uniaxial PVDF film, the performance of all 10 sensing elements is similar when $\mu = 0$. However, in the presence of friction, the outputs of sensing elements are different and depend on the friction coefficient as well as the location of the elements.

3.9 Experiments

In order to validate the results from the simulations of the influence of surface friction on the output of a PVDF film compressed between two plates, a set of experiments were performed. Prior to conducting the main experiment, the friction coefficients of the sample

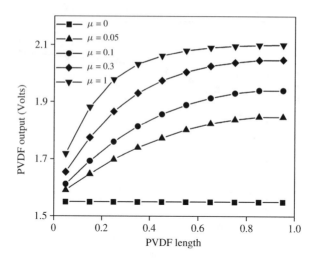

Figure 3.19   The output voltages of 10 equal biaxial PVDF sensing elements evenly positioned along the x-axis. While the applied load is kept constant, the friction coefficient is varied between 0 and 1. The horizontal axis is normalized by half the length of PVDF film, and hence 0 represents the center of the film and 1 denotes the PVDF edge at $x = 15$ mm

surfaces had to be estimated. Once the friction coefficients of the surfaces on both sides of the PVDF film were experimentally determined, the characterized surfaces were used in another experimental setup to record the output voltage of the PVDF film.

3.9.1   Surface Friction Measurement

In order to measure the static friction coefficient of test surfaces, as shown in Figure 3.20, a conventional experimental setup was designed and implemented. Extreme care was taken to reduce unwanted friction forces. The minimum force increment steps were 0.01 N and the force applied to the surfaces was in the range 2–10 N. Some of the selected surfaces were made of standard abrasive papers. In addition, other available surfaces such as stationery paper or Plexiglas were also tested. To achieve low friction surfaces, lubricants were used. However, because of various problems, the application of oils did not yield satisfactory results. The first difficulty was the stickiness of the smooth surfaces in the presence of the oil. The second was maintaining the uniformity of the lubrication conditions, such as the thickness of the oil film present for different applied loads. To eliminate these difficulties, a couple of readily available dry lubricants were tested. Normally these materials can be sprayed onto the target surfaces.

Alternatively, low-friction materials, such as Teflon tape, which can be cut and pasted on the surface, can be used. For the present study, 555 Silicon Dry Film Lubricant and a Molykote 321 Dry Film Lubricant from Dow Corning were used. In addition, a PTFE adhesive tape from Nitto Denko was tested. The data recorded for silicon dry film was not repeatable because the silicon film peeled off easily during the experiments. The selected abrasive papers ranged in grade between very fine (grade 2000) to very rough (grade 240). Since one of the surfaces in the tests was PVDF, always one surface

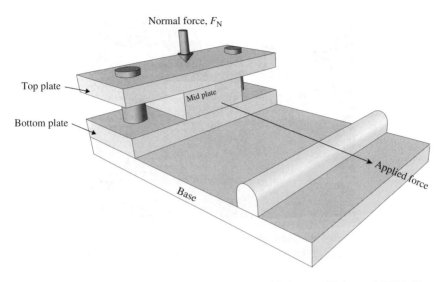

Figure 3.20 The experimental setup used to measure friction coefficients. PVDF films were adhered to both sides of the mid-plate. The friction surfaces were attached to the internal surfaces of the top and bottom plates

Table 3.3 List of abrasive materials and their corresponding friction coefficients

Surface	Black dry film	PTFE tape	Stationery paper	Abrasive 2000	Abrasive 1200	Abrasive 800	Abrasive 240
μ	0.11	0.19	0.29	0.35	0.40	0.44	0.55

was tested against the PVDF film. However, the tested PVDF film was metalized with aluminum so it was this coating that was considered to be the real contact surface in the tests. Table 3.3 shows the tested materials and their corresponding static friction coefficients established through testing.

3.9.2 Experiments Performed on Sandwiched PVDF for Different Surface Roughness

An experimental setup, as illustrated in Figure 3.21, was used to apply a known load to a PVDF film pressed between two friction surfaces. A square-shaped 110 μm-thick PVDF film was placed between two thick plates of Plexiglas and the mentioned pre-characterized surfaces were applied one by one between the plates and the top and bottom surfaces of the PVDF film, as shown in Figure 3.22.

A 20 N load, at a frequency of 20 Hz was applied to the top Plexiglas. Considering the area of the PVDF film (30 mm × 30 mm), the maximum pressure exerted on the film was about 22 kPa. A picture of the experimental setup is shown in Figure 3.23.

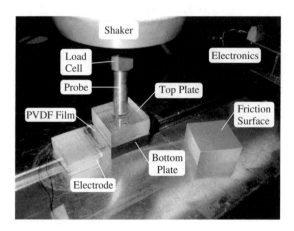

Figure 3.21 In this part of the experimental setup the PVDF film is sandwiched between two plates. The internal surfaces of the plates are covered with the pre-characterized surfaces

Figure 3.22 The configuration used for the experiment to measure the PVDF response to the friction material

To apply a repetitive load, a signal generator with sinusoidal output was used. Before connecting this signal to the shaker, the signals were amplified using a power amplifier (V203, PA25E-CE; Ling Dynamic Systems).

To characterize the output voltage of the PVDF films sandwiched between the two surfaces, a calibrated load sensor (Kistler 9712B50) was used to record the amplitude of the applied load. To record and analyze the output of the PVDF film when the friction coefficient of the surfaces was changed, a data acquisition system (National Instrument, NI PCI-6225) running LabVIEW software was used. To connect the piezoelectric charge output to the DAQ, a charge amplifier was used in which an operational amplifier converts the charge into voltage. In contrast to the voltage-mode amplifiers in which the output voltage depends on the input impedance, the output voltage in charge-mode amplifiers depends on the feedback capacitor, C_f and the charge developed on the piezoelectric film, Q. Therefore, the output voltage of the charge amplifier is independent of the cable capacitance. This is one of the main advantages of using charge amplifiers in piezoelectric applications. The voltage gain can be determined by the ratio Q/Cf. Using a LabVIEW

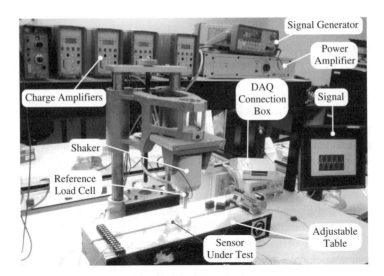

Figure 3.23 A view of experimental setup used to measure the PVDF response to the friction materials

built-in Butterworth low-pass-filter, the 60 Hz line noise was eliminated and the peak-to-peak output, together with the corresponding input, was recorded. This data was then saved in a file and plotted against the theoretical results, as shown in Figure 3.24 in which the theoretical curve is the result of an FE model for the uniaxial PVDF film executed for an applied load of 22 kPa.

Figure 3.24 also shows experimental data that reveals a decaying trend when the friction coefficient is increased. However, for friction coefficients greater than 0.5, the projections of the abrasive papers would have caused permanent and local scratches leading to an increase in output. Therefore, for $\mu > 0.6$, no consistent data was obtained.

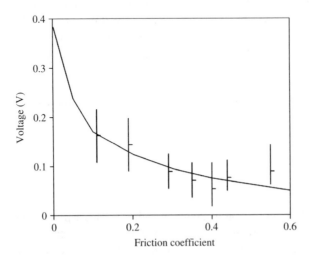

Figure 3.24 Comparison between theoretical data and experimental results. For both cases a 20 N force is applied (error bar indicates the range of readings)

3.10   Discussion and Conclusions

This chapter has dealt with the manner in which the proposed tactile sensor will be developed, based upon prevailing theoretical and experimental knowledge. Because the goal is to utilize the mechanical and electromechanical properties of piezoelectric PVDF in the following designs, the anisotropic behavior of PVDF was the main focus of this study. The result of this study has provided, not only a sound basis for the development of our proposed sensor, but it has also contributed to exploring the piezoelectricity phenomena of PVDF in different configurations, such that they can be used in a variety of applications.

In the theoretical approach, we analyzed the anisotropic behavior of PVDF film using the FE method effect, which consists of measuring the angle between the material and global coordinate systems. In addition, consideration was given as to the difference between uniaxial and biaxial PVDF film, as well as the performance of PVDF film in thickness mode, when the film is sandwiched between two plates. The main concern in this case was the effect of friction on the output of the PVDF. Using ANSYS, an FE contact analysis associated with the modeling of the PVDF piezoelectric was developed and the results for both uniaxial and biaxial PVDF, for a range of friction coefficients, were obtained. It is also shown that the PVDF response is greatly influenced by contact friction. Hence, one of the key reasons for observing contrasting outputs from a structurally similar manufactured sensor could be attributed to the difference in contact friction. Therefore, extreme care should be paid to the uniformity of the contact surfaces during the manufacturing process. Simulations show that when a PVDF film is firmly glued to a surface, it is equivalent to having infinite friction with that surface. This is due to the fact that the boundary condition (e.g., gluing, etc.) not only affects the output of the sensors, but also makes it difficult to manufacture a number of similar sensors.

Another benefit of this study was that it enabled the measurement of coefficient of friction of the contact surfaces. This technique is particularly appropriate for the low-friction range where accurate measurement is difficult and other available friction measurement methods might be inadequate.

In experimental approach, the Young's modulus of PVDF film in directions 1 and 2 were obtained. To investigate the dependency of PVDF output to friction, a number of surfaces were tested in order to determine their coefficients of friction and which were, subsequently, used in another experimental setup in which PVDF film was sandwiched between the characterized surfaces.

In practice, based on what we learned from all the force and tactile sensing applications in this chapter, we recommend that PVDF be used in the following order:

1. Drawn direction
2. Transverse direction
3. Thickness mode.

Although the thickness mode is shown as being the least preferred method, sometimes the type of application prevents the use of either of the other two methods. In such cases, the user is advised to pay attention to the boundary conditions, upon which the output of the sensor is highly dependent.

References

1. Nye, J.F. (1960) *Physical Properties of Crystals*, Oxford University Press.
2. Wang, T.T., Herbert, J.M., and Glass, A.M. (1988) *The Applications of Ferroelectric Polymers*, The Blackie Publishing Group, Glasgow/London.
3. ANSYS Inc. Documentation for ANSYS, Modeling Material Nonlinearities/Mixed u-p Formulation, Release 10.
4. Cuhat, D.R. (1999) Multi-point vibration measurement using PVDF piezoelectric film with application to direct sensing of modes in structures. PhD. Purdue University.
5. Yongrae, R., Varadan, V.V., and Varadan, V.K. (2002) Characterization of all the elastic, dielectric, and piezoelectric constants of uniaxially oriented poled PVDF films. *IEEE Transactions on Ultrasonics, Ferroelectrics and Frequency Control*, **49**, 836–847.
6. Bicchi, A., Canepa, G., Rossi, D.D. *et al.* (1996) A sensorized minimally invasive surgery tool for detecting tissue elastic properties. Proceedings of the 1999 IEEE International Conference on Robotics and Automation, Minneapolis, MN, 1996, pp. 884–888.
7. Dargahi, J. (2002) An endoscopic and robotic tooth-like compliance and roughness tactile sensor. *Journal of Mechanical Engineering Design (ASME)*, **124**, 576–582.
8. Dargahi, J., Parameswaran, M., and Payandeh, S. (2000) A micromachined piezoelectric tactile sensor for an endoscopic grasper-theory, fabrication and experiments. *Journal of Microelectromechanical Systems*, **9**, 329–335.
9. Dario, P. (1991) Tactile sensing: technology and applications. *Sensors and Actuators A: Physical*, **26**, 251–256.
10. Howe, R.D., Peine, W.J., Kantarinis, D.A., and Son, J.S. (1995) Remote palpation technology. *IEEE Engineering in Medicine and Biology Magazine*, **14**, 318–323.
11. Toda, M. and Tosima, S. (2000) Theory of curved, clamped, piezoelectric film, air-borne transducers. *IEEE Transactions on Ultrasonics, Ferroelectrics and Frequency Control*, **47**, 1421–1431.
12. Klaase, P.T.A. and Van Turnhout, J. (1979) Dielectric, pieso- and pyroelectric properties of the three crystalline forms of pvdf. Presented at The 3rd International Conference On Dielectric Materials, Measurements and Applications, Birmingham, UK, 1979.
13. Birlikseven, C., Altinta, E., and Durusoy, H.Z. (2001) A low-temperature pyroelectric study of pvdf thick films. *Journal of Material Science in Electronics*, **12**, 601–603.
14. Royer, D. and Kmetik, V. (1992) Measurement of piezoelectric constants using an optical heterodyne interferometer. *Electronics Letters*, **28**, 1828–1830.
15. Dargahi, J. (1998) Piezoelectric and pyroelectric transient signal analysis for detection of the temperature of a contact object for robotic tactile sensing. *Sensors and Actuators A: Physical*, **71**, 89–97.
16. Zheng, Z., Guy, I.L., and Trevor, L.T. (1998) Piezoelectric coefficient of thin polymer films measured by interferometry. *Journal of Intelligent Material Systems and Structures*, **9**, 69–73.
17. Guy, I.L. and Zheng, Z. (2001) Piezoelectricity and electrostriction in ferroelectric polymers. *Ferroelectrics*, **264**, 33–38.
18. Kepler, R.G. and Anderson, R.A. (1978) Piezoelectricity and pyroelectricity in polyvinylidene fluoride. *Journal of Applied Physics*, **49**, 4490–4494.
19. Goodfellow, Technical Information of Polyvinylidenefluoride (PVDF). Access year: 2007 USA, www.goodfellow.com.
20. Airmar Technology Corporation New hampshire 03055-4613, USA, Technical Information of Polyvinylidenefluoride (PVDF). Access year: 2007 www.airmar.com.
21. Ueberschlag, P. (2001) PVDF piezoelectric polymer. *Sensor Review*, **21**, 118–126.
22. Klaase, P.T.A. and Van Turnhout, J. (1979) Dielectric, pieso- and pyroelectric properties of the three crystalline forms of pvdf. Presented at the 3rd International Conference On Dielectric Materials, Measurements and Applications, Birmingham, UK, 1979.
23. Simo, J.C. and Laursen, T.A. (1992) An augmented lagrangian treatment of contact problems involving friction. *Computers & Structures*, **42**, 97–116.
24. ANSYS Inc. Theory Reference, ANSYS Release 9.0.002114, 2004.

4

Design, Analysis, Fabrication, and Testing of Tactile Sensors

Tactile sensors suitable for endoscopic applications should satisfy many criteria. Among these criteria, the capability to grasp slippery tissue, to determine pressure distribution and to measure the softness of the grasped object seem to be the essential and often challenging ones to address. Present-day endoscopic tools such as graspers have rigid tooth-like features to grasp slippery tissue (see Figure 4.1), therefore any similar device, equipped with tactile sensors, should maintain the capability of grasping slippery tissue reliably. Although several papers have been written on the challenges of designing tactile sensors for minimally invasive surgery (MIS) applications, the proposed designs submitted mainly accommodate forces in the order of grams and, furthermore, are unable to determine pressure distribution [1–9].

This chapter presents the detailed design of two different endoscopic graspers. Section 4.1 describes an endoscopic force sensor which is able to measure the force along its length. Section 4.2 presents the design and prototype modeling of a multi-functional endoscopic grasper. Both devices utilize the piezoelectric polymer polyvinylidene fluoride (PVDF) of which a detailed description is given in Chapter 3. Both devices are designed such that they can be manufactured using conventional microelectromechanical systems (MEMS) techniques.

MEMS devices offer competitive advantages due to their batch fabrication capabilities, small size, improved functionality, and low cost, due to integrated circuitry (IC). The incorporation of MEMS devices in surgical tools represents one of the greatest growing areas of improvement in the medical sector. Not only does MEMS technology improve surgical outcomes by reducing the incumbent risk attached to such procedures, it also ameliorates costs by providing surgeons with real-time data about instrument force, performance, tissue density, and temperature, as well as providing better and faster methods of tissue preparation, grasping, cutting, and extraction [10].

This book has, thus far, elaborated upon the multi-functional aspects of the tactile sensor, together with the working principles and characteristics of piezoelectric PVDF films as transducers. The ultimate objective, however, is to integrate this proposed tactile sensor with existing MIS graspers, although in order to accomplish this, the current sensor must be miniaturized.

Tactile Sensing and Displays: Haptic Feedback for Minimally Invasive Surgery and Robotics, First Edition.
Saeed Sokhanvar, Javad Dargahi, Siamak Najarian, and Siamak Arbatani.
© 2013 John Wiley & Sons, Ltd. Published 2013 by John Wiley & Sons, Ltd.

4.1 Endoscopic Force Sensor: Sensor Design

The structure of the sensor explained in this section consists of three layers. The top layer is made of micro-machined silicon, which has a rigid tooth-like structure similar to that shown in Figure 4.1. The bottom layer of the grasper consists of a flat Plexiglas substrate. Sandwiched between the Plexiglas and silicon is a 25 μm thick PVDF film. The two flat surfaces of the PVDF film are patterned aluminum electrodes that provide electrical output to an amplifier. The top side consists of four strips of aluminum electrode and the bottom side has a single common electrode covering the entire area occupied by the top four electrode strips. The areas of intersection between the top and bottom electrodes form the active piezoelectric sensors. Four output signals are derived from the device. The top exposed portion of the silicon is anisotropically etched to form four tooth-like structures at 3 mm intervals. A 6 mm thick Plexiglas plate was prepared as a substrate for the sensor. The PVDF film is attached between the silicon and the substrate by employing electrically nonconductive glue in such a way that the four sensing elements on the PVDF film are firmly affixed directly below the four tooth-like structures on the silicon. A detailed drawing of the device is shown in Figure 4.2. A cross-sectional view of the sensor is illustrated in Figure 4.3. When force is applied to any point on the surface of the silicon, the stress in the PVDF results in a polarization charge at each surface. The amplitude of the signal is proportional to the magnitude of the applied force, and differences between the slopes of the four output signals indicate any localized position of the applied force.

4.1.1 Modeling

The silicon is modeled as a uniform beam of finite length located on an elastic foundation (such as PVDF/Plexiglas substrate). When a concentrated load P (point force) is applied at $x = a$ (e.g., point C in Figure 4.4) on a beam, the lateral load is transmitted to the substrate by the shear force which, at any point, is given in [12].

$$EI\frac{d^4v}{dx^4} = Q \tag{4.1}$$

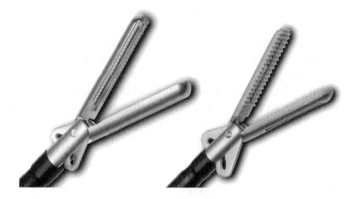

Figure 4.1   Endoscopic graspers [11] (© Microline Surgical)

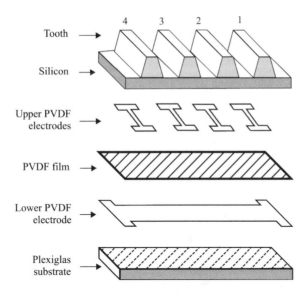

Figure 4.2   Expanded view of the tactile sensor unit

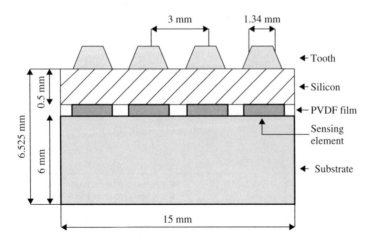

Figure 4.3   Cross section of the tactile sensor unit

where Q = shear force, EI = flexural rigidity (bending modulus) of the beam, and v is the deflection along the y-axis of the beam at position x of the applied load.

For a given magnitude and position of the applied load on the beam, an expression for the shear force can be obtained from [12]:

$$Q = \frac{P}{\sin h^2 (\lambda l) - \sin^2 (\lambda l)} \{(\sin h \, (\lambda x) \sin (\lambda x) + \sin h \, (\lambda x) \cos (\lambda x))\}$$
$$. \, (\sin h \, (\lambda l) \cos (\lambda a) \sin h \, (\lambda b) - \sin (\lambda l) \cos h \, (\lambda a) \cos(\lambda b)$$

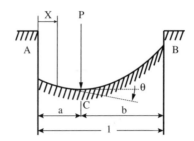

Figure 4.4   Beam on an elastic foundation for theoretical analysis of deflection. Load P is applied at the location, θ is the slope of the beam at the point of application of the load

$$+ \sin h\,(\lambda x)\,\sin\,(\lambda x)\,.[\sinh\,(\lambda l)\,(\sin\,(\lambda a)\cos h\,(\lambda b) - \cos\,(\lambda a)\sin h\,(\lambda b))$$

$$+ \sin(\lambda l)(\sin h\,(\lambda a)\cos\,(\lambda b) - \cos h\,(\lambda a)\sin(\lambda b))]) \qquad (4.2)$$

where P = applied force, $\lambda = 4\sqrt{K/4EI}$, $K = \omega K_0$, K_0 = foundation modulus in N mm^{-3}, and ω = width of the silicon in contact with the PVDF/Plexiglas substrate.

Figure 4.5 shows the distribution of shear force Q along the length of the sensor structure (i.e., total length of the sensor, l = 15 mm) for a concentrated load applied at locations x = 3 and 6 mm away from the edge of the silicon (i.e., tooth numbers 1, called Touch 1 and tooth No 2 called Touch 2 in Figure 4.2). The analytical results confirm the intuitive observation that, as the distance from the concentrated load increases, the shear force decreases.

The slope angle θ, which defines the angle between the tangent to the deflected curve at the load point and reference axis, is given in [12]:

$$\theta = \frac{2P\lambda^2}{K}\frac{\cosh^2\,(\lambda a)\cos^2\,(\lambda b) - \cos^2(\lambda a)\cosh^2(\lambda b)}{\sinh^2\,(\lambda l) - \sin^2(\lambda l)} \qquad (4.3)$$

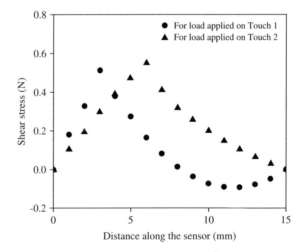

Figure 4.5   Theoretically derived stress distribution along the sensor length for a point load

In both Equations 4.1 and 4.2, the responses are for the segment AC of the beam where $x < a$, (see Figure 4.4). The same approach is used for segment BC of the beam where $a < x < l$.

4.1.2   Sensor Fabrication

Sensor fabrication employs, to some extent, procedures that are currently employed in process technology used for semiconductor device fabrication, deposition of material layers, patterning by photolithogrophy, and etching to produce required shapes and patterns.

4.1.2.1   Fabrication of Silicon Part

This sequence is illustrated in Figure 4.6. A (100) double-sided polished silicon wafer was first RCA cleaned and oxidized (0.5 µm thick thermal oxide). A set of four square patterns was lithographically transferred [13] to one side of the silicon wafer (side b), which served as alignment marks for positioning the PVDF film. A buffered oxide etchant (BOE) was used to remove the oxide, after which the exposed alignment pattern was then etched anisotropically (using PSE-300) for a short period in order to create the alignment indentations (see Figure 4.6-i).

   The wafer was then stripped of all oxide using buffered oxide etch (BOE), RCA cleaning, and re-oxidization. Using a double-sided mask aligner, the tooth-like pattern was lithographically transferred onto the side of the silicon wafer. After patterning the oxide using BOE, the wafer was etched anisotropically to form the tooth-like structure (see

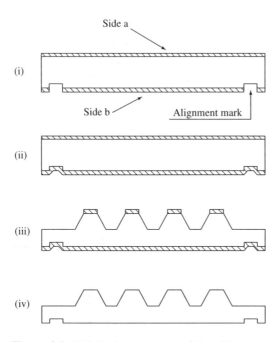

Figure 4.6   Fabrication sequence of the silicon part

Figure 4.6 ii, iii). Appropriate corner compensation was given to the edges of the structure so that the undercutting of the anisotropic etchant did not affect the tooth-like pattern due to the long etch duration of this step. The cross section of the final silicon part is shown in Figure 4.6 iv.

4.1.2.2   Patterning the PVDF Film

A 25 μm uniaxially drawn, metalized (aluminum) and poled PVDF film was cut as a 20×14 mm section. This dimension was 5 mm larger on each side than the footprint of the silicon piece. The piezoelectric coefficients of the PVDF film, d_{31}, d_{32}, and d_{33} were 20, 2, and -20 pC N^{-1}, respectively. The PVDF sample was carefully taped onto a clean silicon wafer (dummy wafer) such that the film was stretched uniformly over the wafer, and the tape covered all sides, overlapping by at least 2 mm. Using a spinner, photoresist was spun onto the PVDF film. In order to ensure a uniform coating, the spin speed was slowly increased from 0 to 500 RPM. The resultant film was then soft baked at 50 °C for 40 minutes. This process was repeated for the second side of the PVDF film. The masks containing the patterns for the two sides of the PVDF film were aligned and held on top of each other. The photoresist-coated PVDF film was then inserted between the pre-aligned masked plates. Using a UV source for the mask alignment, both sides of the photoresist-coated PVDF film were exposed consecutively. The exposed film was then developed, rinsed, and baked at 50 °C for 60 minutes.

The electrodes were patterned using a commercial aluminum etchant. Acetone was used to remove the photoresistive material. The PVDF film was rinsed in de-ionized (DI) water and later dried. Figure 4.7 illustrates the steps for the PVDF electrode pattern process.

4.1.2.3   Sensor Assembly

A thin layer of nonconductive epoxy was applied to the underside of the silicon wafer as shown in Figure 4.7. Using a stereo microscope, the PVDF film was aligned using the four alignment marks, and attached to the silicon. This assembly was uniformly pressed

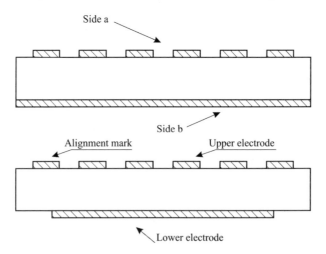

Figure 4.7   Fabrication steps of the PVDF film

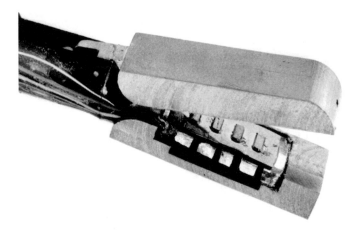

Figure 4.8 Photograph of the prototyped endoscopic grasper integrated with the tactile sensor (See Plate 1)

using a compression block to ensure even adhesion, after which it was cured inside an oven at 50°C for a period of 1 hour. A 6 mm thick Plexiglas substrate of 25 × 45 mm dimensions was prepared, and a uniform layer of nonconductive glue was applied. The Plexiglas substrate was glued to the PVDF film such that the PVDF film was sandwiched between the Plexiglas and the silicon. The whole assembly was compressed using a compression block and cured at 50 °C for 1 hour. An illustration of this miniaturized sensor integrated onto an endoscopic grasper is shown in Figure 4.8.

4.1.3 Experimental Analysis

An experimental set-up was developed to analyze the performance of the sensor (see Figure 4.9). A 2 mm diameter circular probe, which was driven by a vibration unit and actuated by a 15 Hz sinusoidal signal, was used to apply a force. The charge generated by the four piezosensitive regions was amplified using a charge amplifier and the output was measured using an oscilloscope. The magnitude of the applied force was determined by a force transducer inserted between the probe and the vibration unit.

To avoid damage to the sensor, the maximum applied load was limited to 2 N. Figure 4.10 shows the typical response of each PVDF sensor, which indicates that the

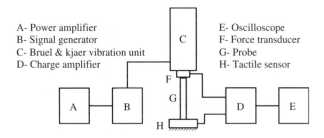

A- Power amplifier
B- Signal generator
C- Bruel & kjaer vibration unit
D- Charge amplifier

E- Oscilloscope
F- Force transducer
G- Probe
H- Tactile sensor

Figure 4.9 Experimental measurement set-up

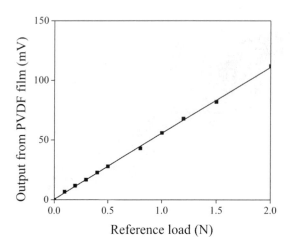

Figure 4.10 Response of the PVDF sensor for linear load input

output voltage from each varies linearly with the applied force. A second set of experiments was conducted by applying an oscillatory force (1 N maximum) at 15 Hz to the center of each tooth, and the output voltage from each individual sensing element was recorded. Figure 4.11 shows the results of these experiments in which it can be seen that when a force is applied to a tooth, the distribution of shear force (which is proportional to the measured voltage) is almost linear and away from the location of the applied load.

When a sinusoidal force is applied, the output charge from each PVDF sensing element is the sum of piezoelectric coefficients, d_{31}, d_{32} and d_{33}, multiplied by the magnitude of the applied force (i.e., $P = q/(\psi_1 d_{31} + \psi_2 d_{32} + d_{33})$), where q is the output charge, P is

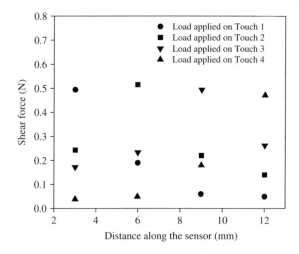

Figure 4.11 Measured shear stress distribution along the length of the sensor for a 1 N applied force at Touch 1 to Touch 4, respectively

the applied force on a sensing element, ψ_1, ψ_2 are constants proportional to the electrode area of the sensing elements, and d_{31}, d_{32}, d_{33} are piezoelectric coefficients in the drawn, transverse, and thickness directions [14]. Therefore, the output voltage values are given by:

$$V = P/C \left(\psi_1 d_{31} + \psi_2 d_{32} + d_{33} \right)$$

where C is the capacitance of the PVDF film. Due to the constrained configuration of the PVDF (i.e., glued to both silicon and Plexiglas), the exact contribution of each piezoelectric coefficient in the output voltage is difficult to determine (see Chapter 3 for detailed information). Thus, the magnitude of the shear force from the output charge, due to each of the sensing elements, is difficult to determine [14]. However, to compare the theoretical shear stress values with the experimental values for a given set of tests (i.e., same applied load), the ratios of the theoretical shear force and output voltage from each sensing element were plotted against the distance along the sensor. The result shows that the difference in the theoretical values compared to the experimental values for each sensing elements was less than 20%. By using the least squares method, an estimation of the above constant (i.e., $\psi_1 d_{31} + \psi_2 d_{32} + d_{33}$) was obtained. The error in the experiment was also estimated using this method. By multiplying this constant by the output voltages from each sensing element, the experimental shear force was obtained. These values were then plotted against the theoretical values for comparison.

The comparison of the theoretical and typical experimental values for each sensing element is shown in Figure 4.12a–d. It can be seen that the shear force decreases as the distance from the center of the applied force increases. The stress reduces exponentially from the sensing element, as shown in Equation 4.1 (see theoretical analysis). The difference between the experimental and theoretical results was determined to be less than 20%.

Equation 4.3 shows that the slope varies as a function of the location of the applied force (i.e., parameters 'a' and 'b'). For example, the magnitude of the slope when a force of 1 N was applied on tooth numbers 1 and 2 (and also for tooth numbers 3 and 4) was calculated, and the values for the slope were 1.37×10^{-11} and 1.19×10^{-13}, respectively. The magnitude of the slope changes by varying the position of the applied force on the sensor. The slope at the location of the applied force is used to determine the location of the load on the sensor.

As previously mentioned, when a force is applied to the tactile sensor unit, bending stress occurs in the silicon. This stress might have the effect of pulling the PVDF film in the transverse direction (perpendicular to the applied force), resulting in an additional output charge, due to the shear stress. However, since the difference between the modulus of elasticity of the silicon and elastic foundation PVDF/Plexiglas is negligible (i.e., only two orders of magnitude) and the magnitude of the applied force is relatively small (2 N max), the bending stress would be small in comparison to the magnitude of the shear stress. Hence, this effect is ignored in the results.

In the design of the sensor, care has been taken to avoid the pyroelectric effect. Since the PVDF film was sandwiched between the silicon and Plexiglas, there is sufficient thermal insulation to reduce this effect and, to reduce it even further during experimental testing, a thin layer of Mylar film was placed between the probe and silicon. This precaution had the additional advantage of avoiding any damage to the silicon by a possible impact from the probe.

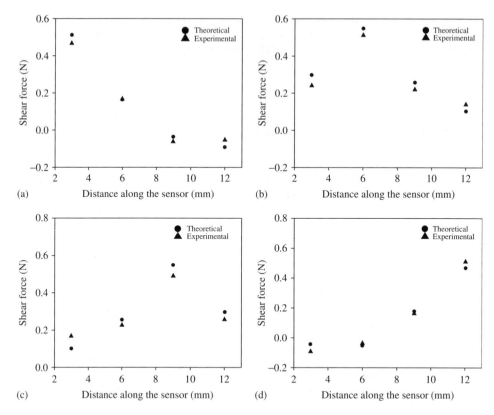

Figure 4.12   (a) Comparison between theoretical and experimental stress distribution for 1 N applied force, concentrated load applied on Tooth 1. (b) Comparison between theoretical and experimental stress distribution for 1 N applied force, concentrated load applied on Tooth 2. (c) Comparison between theoretical and experimental stress distribution for 1 N applied force, concentrated load applied on Tooth 3. (d) Comparison between theoretical and experimental stress distribution for 1 N applied force, concentrated load applied on Tooth 4

Since the silicon structure is glued to a rigid base, it can withstand a very high compressive stress; however, the teeth of the sensor may be damaged due to the application of shear stress along the sensor while handling tissues. This limitation could be reduced by proper encapsulation of the sensor, (e.g., providing a thin layer of appropriate tissue-compatible polymer coating). This approach could also reduce the shear force effect. One of the major limitations of the reported sensor is the lack of DC response due to the nature of the PVDF piezoelectric film. This limitation could, however, be overcome by applying a step input force, and the peak value of the transient signals could be captured to determine the instantaneous output [15]. In reality, the grasping action of a tissue is close to either a step or a sinusoidal load (see Chapter 5).

The sensor presented here is fairly simple to fabricate and assemble. Compared to other reported sensor arrays, this design uses only four channels, thereby reducing cross talk problems when acquiring and processing tactile information. Due to the use of PVDF film, this sensor exhibits high sensitivity, a large dynamic range, a wide bandwidth with good linearity, and a high signal-to-noise ratio [5].

Using the design demonstrated here, for a concentrated load, the entire surface of the sensor can be used to sense the force as well as its position. This effect is very novel compared to other reported tactile sensor arrays where the region between the two sensing elements is essentially a dead area (no sensitivity) [16]. The magnitude of the applied force can be obtained from the magnitude of the output from PVDF sensing elements, and the position of the applied force can be found from the value of the slope at the point of application of the load. This information can be displayed, or fed back to the surgeon's finger, by means of a haptic interface.

A more sophisticated multi-functional sensor system is described below. Although this sensor is capable of measuring the force, and the position of the force, both along the grasper length as well as its breath, it was mainly designed to measure the softness of the grasped object. It is shown that a wide range of soft objects, including soft tissues, can be differentiated by this sensor. There are other features, such as lump detection which are discussed in detail later in this chapter.

4.2 Multi-Functional MEMS–Based Tactile Sensor: Design, Analysis, Fabrication, and Testing

This section describes the design, microfabrication, and testing of a second endoscopic tactile sensor whose design and geometry are explained in the following subsections. A finite element model of the sensor is also developed. To check the accuracy of the numerical model, its deflection profile for a given load is compared with that of a closed-form relationship. The microfabrication steps are also explained and then the experimental results for softness sensing are compared with the results of this finite element model.

4.2.1 Sensor Design

An illustration of the complete grasper (host device) is shown in Figure 4.13, in which either the lower or upper jaw is equipped with an array of sensor units. Although the number of teeth may vary in different graspers and/or applications, this modular design

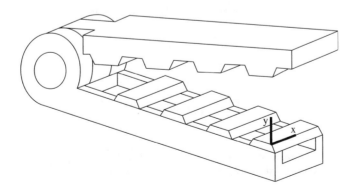

Figure 4.13 The MIS grasper equipped with an array of the proposed sensor

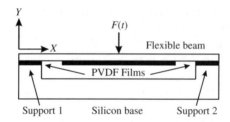

Figure 4.14   Cross section of the sensor unit (one tooth)

assures that the performance of all sensor units is essentially similar. Therefore, the operation of this array can be studied and analyzed by considering a single sensor unit and only one tooth of the grasper.

The cross section of the sensor unit subjected to a time-dependent force $F(t)$ is shown in Figure 4.14. In Figures 4.13 and 4.14, the X and Y directions are selected, similar to the traditional notation in classical mechanics. In order to find the softness of the grasped object, both the amplitude of the applied load and the resultant deflection are required simultaneously. The sensing mechanism utilizes PVDF films, both in the thickness mode at the end supports to measure the applied load, and in extensional mode attached to the flexible beam to measure the bending stresses. Knowing the applied load from the PVDF films at the supports and the amount of deflection or the equivalently developed bending stresses from the middle PVDF film, softness characterization of the objects is feasible. The concept of the sensor operation is depicted in Figure 4.15. As can be seen

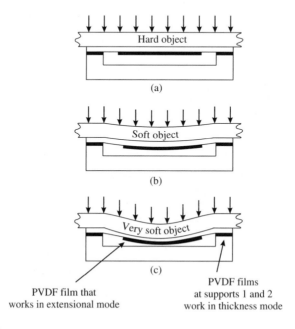

Figure 4.15   The basic scheme for the proposed softness sensing technique

in Figure 4.15a, when a rigid object is used as the target, no deformation occurs in the beam. Although the PVDF films at the supports respond to the applied force, no output from the middle PVDF film (extensional mode) is expected. However, a soft object under a load F will bend the beam.

As shown in Figure 4.15b, c, larger beam deflection occurs for softer contact objects. Therefore, the PVDF film, which is firmly adhered to the beam, elongates, and develops a charge proportional to the bending stress which, itself, is dependent on the extent of bending. The maximum amount of bending can be controlled at the design stage by selecting the material of the flexible beam, as well as its thickness and length, for a maximum given load. A beam with higher Young's modulus, or thickness, shows less deflection, but can sustain a higher applied load. Conversely, a soft beam deflects easily and produces a higher output charge, but it may not be a good choice for high load applications since it cannot develop enough gripping load. Thus, with such a design, a balance between the softness, load, and the design output charge must first be established. For a given sensor, the thickness and material property of the beam are known, thus the deflection of the beam will only be a function of the flexural rigidity of the contact object.

As described, the sensor consists of a top and a bottom part. The bottom part forms the supports and is the base for the top part which consists of hanging beams. The top part encompasses the clamped–clamped hanging beams which are used for softness measurements. For softness sensing, the applied force, as well as the resultant deformation, needs to be recorded. The magnitude of the applied load can be measured by piezoelectric films that are sandwiched between the top and bottom parts. However, to measure the resulting deflection, PVDF films are attached to the beams. Figure 4.16 shows how this sensor can be integrated with MIS graspers.

A sensor unit, for instance, with two or more beams can be considered as a module. In the remainder of this chapter, to avoid unnecessary complexity, a design with three teeth, as shown in Figure 4.17, is considered. This module can then be repeated in an array to cover the whole length (or width) of the grasper. Figure 4.17 shows the backside view of the top part of a sensing module with three sensing elements.

Three PVDF films working in extensional mode (d_{31}) were attached to the beams in order to measure the deflection. Two end support PVDF films, working in the thickness

Figure 4.16 A proposed design of the smart grasper in which the incorporation of a microfabricated sensor is presented (See Plate 2)

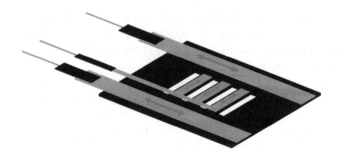

Figure 4.17 Backside view of the sensor's top part. The positions of the PVDF films and the electrode configurations are illustrated in this view. Arrows on the PVDF film strips indicate the orientation of the drawn direction of the PVDF film (See Plate 3)

mode (d_{33}), were sandwiched between the top and bottom parts in order to measure the total applied load.

To ensure a firm grasp, the ideal sensor should be corrugated, which has been taken into consideration for the current design. Depending on the application, it is possible to have one sensorized jaw or to have both jaws sensorized. In the latter case, the capability of the sensor for locating hidden lumps will increase. For the present study, one jaw was considered for sensor integration.

The dimensions of the sensor, and particularly the beams, should be determined for a given range of force and softness for any specific application. Using a number of parameters, the desired working range or sensitivity of the sensor can be achieved. For instance, the length, width and thickness of the beam can affect the load capacity of the sensor as well as its sensitivity. Short and thick beams withstand higher loads, but show less flexibility and, as a result, lower piezoelectric output would be obtained. Thin, long beams show the opposite behavior and are appropriate for delicate applications such as pulse detection, therefore, depending upon the application, dimensions should be determined accordingly. Another influential factor in determining the overall dimensions of the sensor is the dimensions of the original grasper into which the sensor is to be integrated. The number of hanging beams, or sensing units, should also be determined in order to satisfy the required spatial resolution.

In order to miniaturize a suitable sensor for integration with existing MIS graspers, batch processing techniques should be utilized instead of manual techniques. For instance, PVDF films should be deposited directly onto the target areas.

However, since at the time of prototyping, this technology was not available for microfabricating the sensor, the dimensions were selected in such a way that pre-fabricated PVDF films could be cut and placed on the designated areas. Figure 4.18 shows the dimensions that were considered for this work. The length and width of each beam is 7 and 2 mm, respectively.

Although a sensor with these dimensions cannot be integrated into current MIS graspers, the feasibility of sensor microfabrication, taking into account incumbent difficulties, and sensor capabilities, were investigated in this work. To have maximum sensitivity and yet avoid the difficulties of working with very thin silicon wafers, the thickness of silicon used was 180 µm.

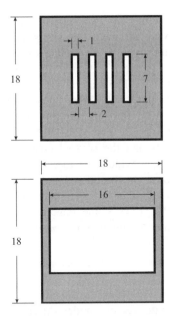

Figure 4.18 The scheme of top and bottom parts (dimensions are in mm)

The main advantage of using an array of sensing modules (instead of a single module) is to register the stress distribution. The profile of stress distribution is especially meaningful when a disruptive object, such as a lump, is embedded in bulky soft tissue.

4.2.2 Finite Element Modeling

Before performing experiments on the sensor, a finite element modeling of MEMS sensor was carried out. The results of the simulation helped to understand the behavior of the sensor under given loads and also the PVDF film response to the loadings. Figure 4.19 shows a meshed model of the sensor. In order to reduce the number of elements, the

Figure 4.19 The meshed FEM model of the MEMS tactile sensor

bottom plate was substituted with only two supports. Since the bottom surface of the bottom part was mechanically constrained, this simplification did not affect the results. This model consisted of two end supports, a top plate with three hanging beams and PVDF films, which were sandwiched between the supports and top plate and also attached to the beams, as shown in Figure 4.17. The material properties of 180 μm thick silicon were considered to be $E = 130$ GPa and $v = 0.28$ for the supports and top plate. The material properties of the 28 μm thick PVDF films used in this model were taken from Goodfellow [17] and are summarized in Table 3.2. The SOLID92 and SOLID98 elements were used for the structural (silicon) parts and the piezoelectric films, respectively. SOLID92 is a 3D 10 node tetrahedral element which has three DOFs at each node: translations in the nodal x, y, and z directions. On the other hand, SOLID98 is a tetrahedral element with a quadratic displacement behavior and is defined by 10 nodes with up to six DOFs at each node. When KEYOPT(1)[1] $= 3$, the degrees of freedom are U_x, U_y, U_z, and VOLT. The governing constitutive equations for piezoelectric materials are described in Chapter 3.

4.2.2.1 Simulation Results

To assess the accuracy of the finite element model, the response of the FEM was compared to that of existing closed form relationship for clamped–clamped beams. As illustrated in Figure 4.20, a distributed load was applied to the first beam and its two supports. The deflection curve, due to the distributed load of 111 kPa, is shown in Figure 4.21.

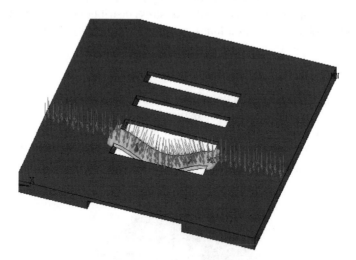

Figure 4.20 The deflected structure when a distributed load of 111 kPa was applied to the first beam and its two supports (See Plate 4)

---

[1] Some ANSYS elements have additional options, known as KEYOPTs and are referred to as KEYOPT(1), KEYOPT(2), and so on. For instance, here KEYOPT(1) = 3 sets the type and number of degrees of freedom of the element. KEYOPTs can be specified using the ET command or the KEYOPT command.

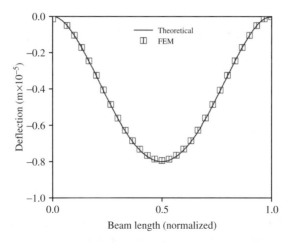

Figure 4.21 The deflection curve of the first beam when a distributed load of 111 kPa is applied to the beam and its two end supports. Solid line illustrates the result obtained from Equation 4.4, while the squares indicate the FEM results

This figure also illustrates the result of theoretical analysis for a clamped–clamped beam under a distributed load, which was obtained from the known formula:

$$y = \frac{1}{24}\frac{Wx^2}{EIL}\left(2Lx - L^2 - x^2\right) \tag{4.4}$$

in which,

L : Beam length (m)

W : Total applied load (N)

E : Young's modulus of the beam (N/m^2)

I : Area moment of inertia (m^4)

x : Distance of point on the beam from the origin, here the left support (m).

It is worth noting that the cross section of the beams after microfabrication using anisotropic wet etching was trapezoidal. In order to compare the experimental results with those of the closed form and FEM, the equivalent rectangular cross section was modeled. It was found that the moment of inertia of a rectangular cross section with a width of 2.12 mm and a thickness of 180 μm is equal to the moment of inertia of the etched cross section. This is shown in Figure 4.22.

This equivalent moment of inertia and cross section were used in the closed form relationship and finite element model. For a distributed load of 111 kPa, the output voltage of PVDF films associated with the first beam was calculated. The output voltage of the middle PVDF film attached to the beam was 680 mV. The calculated voltage for two supports was 85 mV. The equality of voltages of the two supports confirms that the applied load was uniform. For this loading condition, no output voltage from the middle PVDF

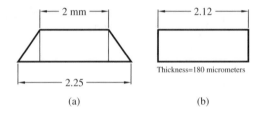

(a)   (b)

Figure 4.22 (a) The actual cross section after anisotropic etching, (b) the equivalent rectangular cross section used in closed form formula and finite element analysis. In both cases the thickness was 180 μm

films associated with the second or third beam was observed. In other words, simulation results showed no cross talk between sensor units due to mechanical load transferring.

In the next simulation, a soft material was considered on top of the first beam and a distributed load was applied to the object. The Young's modulus of the target object was then varied between 10 kPa and 1 MPa. Figure 4.23 shows the modeling geometry and Table 4.1 summarizes the obtained results.

Table 4.1 shows that the maximum deflection varies from 1 to 20.4 μm for a range of Young's moduli between 10 kPa and 1 MPa. The maximum σ_x and the corresponding PVDF output are also shown in this table.

The relationship between the Young's modulus of soft objects and the PVDF output is shown in Figure 4.24.

4.2.3   Sensor Fabrication

In order to fabricate the sensor, several process steps were used. First, the top and bottom silicon parts had to be micromachined. The PVDF films were cut into the designed

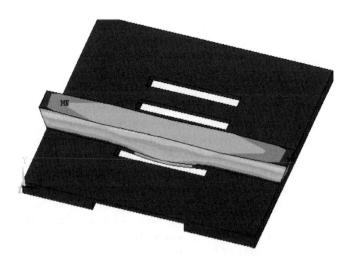

Figure 4.23 The finite element model of the sensor when a soft material is placed on the first tooth and a uniform compressive load is applied (See Plate 5)

Table 4.1 The maximum beam deflection at the center, Y_{max}, the corresponding stress in the x-direction, σ_x, and the output voltage of the middle PVDF film, V_{mid} when the Young's modulus of the contact object, E_{obj} is varied between 0.01 and 1 MPa

E_{obj} (MPa)	Y_{max} (μm)	σ_x (MPa)	V_{mid} (mV)
0.01	−20.4	78.6	170
0.05	−11.2	39.6	95.2
0.1	−7.2	22.8	61.2
0.5	−1.8	4.6	15.3
1	−1.0	2.4	8.5

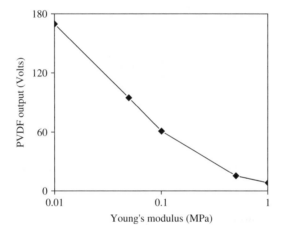

Figure 4.24 The output of the middle PVDF film at constant pressure 111 kPa when the Young's modulus of the contact object is varied between 0.01 and 1 MPa

dimensions and attached in their positions. This section describes the different steps used in the manufacturing of the sensor.

Both silicon parts of the sensor were micromachined using anisotropic wet etching. In order to increase the sensitivity of the sensor, the thickness of the beams was considered to be less than 200 μm. As conventional 500 μm silicon wafer would require an extra 300 μm back etching, a 180 μm thick, 4 inch (100 mm) silicon wafer was used for the fabrication. Several designs with different numbers of beams were considered. The bottom part, however, was the same for all designs. Figure 4.25 shows the steps in the fabrication procedure. A brief explanation of each step of silicon fabrication is schematically shown in Figure 4.26. For simplicity, the cross section of one beam only is shown. The same procedure was used for both silicon parts, however, only the tetramethylammonium hydroxide (TMAH) etching (step 9) time was different for bottom and top parts. More explanation on each step is given below. As the first step, the silicon wafer was diced into the required dimensions (18 × 18 mm).

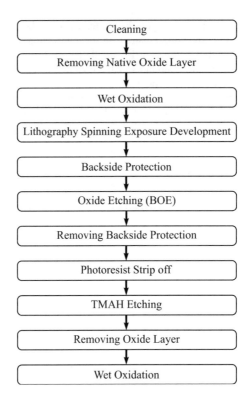

Figure 4.25   Process flow of micromachining process used for top and bottom parts

4.2.3.1   Primary Cleaning

As the first step in the fabrication process, silicon parts were cleaned in order to ensure formation of a homogenous oxide layer. To remove any contamination from the silicon parts, a primary cleansing was performed, mainly for the purpose of removing grease and dust.

4.2.3.2   Pre-cleaning

The samples were placed in a sample holder and immersed in Acetone $((CH_3)_2CO)$ for 10 minutes and then cleaned with isopropanol $(CH_3CHOH–CH_3)$ for 10 minutes. To accelerate the cleaning process, ultrasonic cleaning was applied at 20 kHz. The samples were then rinsed with DI water and afterwards dried using compressed nitrogen gas.

4.2.3.3   Cleaning

To remove any contamination that could still have remained on the surface of the samples, a commonly used Piranah solution (also known as Caro's acid or sulfuric peroxide) [18] was used. Piranah is a solution of two parts by volume of sulfuric acid (H_2SO_4) and one

▬ Silicon Oxide
▬ Photo Resist
▧ Protection Layer

Step	Description	Specification	Schematic
1-1	Pre-Cleaning	10 min in acetone (in ultrasonic bath) 10 min in iso-propanol (in ultrasonic bath)	
1-2	Cleaning	10 min @ Piranah	
2	Removing Oxide Layer	20 Sec @ HF, 1%	
3	Wet Oxidation	20 min for 1000 Å	
4-1	Spin Coating of Photoresist	30 sec @ 4000 rpm	
4-2	Soft Baking	1 min @ 115 °C	
4-3	Exposure	5 sec, a 270 watts mercury lamp	
4-4	Developing	1 min	
4-5	Hard Baking	30 min @ 105 °C	
5	Backside Protection	Using special scoth tape	
6	Oxide Layer Etching	2 min in BOE, 7:1	
7	Removing Backside Protection		
8	Photoresist Strip off	10 min in acetone (in ultrasonic bath) 10 min in iso-propanol (in ultrasonic bath)	
9	TMAH etching	1.5 hr for bottom layer (TMAH 25% @90 °C) 4.5 hr for top layer (TMAH 25% @90 °C)	
10	Removing Oxide Layer	3 min in buffered HF @ room temperature	
11	Wet Oxidation	20 min for 1000 Å	

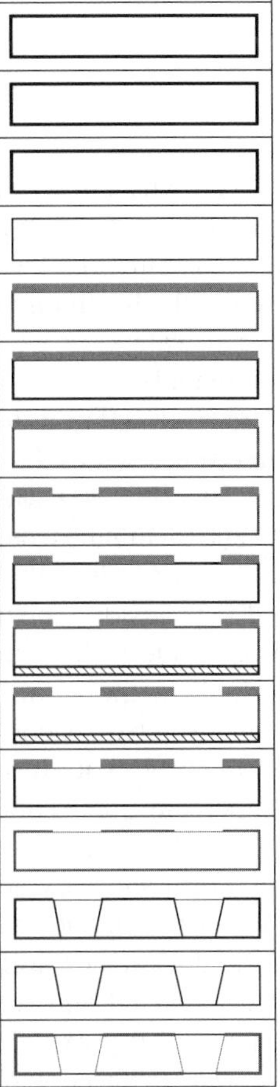

Figure 4.26 The sequence of micromachining process summarized (See Plate 6)

part by volume of 30% hydrogen peroxide (H_2O_2). This solution is exothermic and besides using safety equipment, extra caution in working with hot containers was exercised. Using a sample holder, samples were put into the solution for 10 minutes.

Due to lightness of the samples, this process needed extra attention because the bubbles produced during the reaction stuck to the surface of the samples and made them float. Therefore, a special cage-type sample holder had to be used. After this step, the samples were rinsed in DI water and then dried in nitrogen. Cleaning in Piranah solution creates a thin layer of oxide on the surface of silicon, which can be removed by dipping the samples into a solution of 1% hydrofluoric acid (HF) for 20 s.

4.2.3.4 Oxidation

After this cleaning process, a layer of thermal silicon oxide was grown on the surface of the (100) single crystal silicon samples that served as the mask for subsequent anisotropic etching. Thermal oxidation is a high-temperature process that can be accelerated by using water vapor. The following reactions take place at the silicon surface:

$$Si+O_2 \rightarrow SiO_2 \tag{4.5}$$

$$Si+2H_2O \rightarrow SiO_2 + 2H_2 \tag{4.6}$$

Thermally grown SiO_2 adheres firmly to the silicon substrate without producing either cracks or pores. A furnace (Model RCA from Thermco) was used for this wet oxidation process. Although the main oxidation was mainly wet, short periods of dry oxidation were used at the beginning and end of the process. For annealing and oxidation of these 4 inch silicon wafers, as well as smaller silicon samples, three furnaces were used and stacked horizontally. First of all, the furnace set point was increased to the desired temperature, 1100 °C. While the temperature rose, the bubbler was prepared. The bubbler was rinsed three times with DI water and then filled to approximately two-thirds full with DI water and 10 vol% HCl was then added in order to accelerate the oxidation process. Between 10 and15 minutes before loading the samples into the furnace, the temperature of the bubbler was set to 95 °C and its heater switched on. The samples were put in a glass rack and kept inside the furnace tube. The first sample is usually a dummy sample because, since it faces the stream flow, the silicon oxide obtained might not be as uniform as desired. In the warm-up period, N_2 was connected to the chamber in order to prevent any unwanted reaction during the warm up. This flow also cleanses the chamber and helps create a uniform heat distribution by creating circulation. After reaching the set temperature, the oxidation process was started by disconnecting the N_2 and connecting dry O_2. The dry oxygen flow was continued for 5 minutes because dry oxidation creates a uniform and very adhesive sublayer. Then the oxygen flow was redirected through the bubbler which had reached boiling point and had begun to evaporate. To achieve 1000 Å oxide thicknesses, a 20 minute period of wet oxidation was sufficient. The bubbler was then bypassed and dry oxygen was again connected directly into the chamber. After 5 minutes of dry oxidation, the furnace was turned off. In the cooling-down phase, oxygen flow was replaced with N_2. In order to prevent any thermal shock, samples were gradually removed from the furnace at about 400 °C.

4.2.3.5 Lithography

Patterning on the silicon oxide layer was carried out through a standard lithography procedure, including photoresist spinning, soft baking, exposing the medium using ultraviolet light, development, and finally hard baking. The designed layout of the mask that was used for the patterning is shown in Figure 4.27.

Spin Coating

In order to apply a uniform positive photo resist layer with a thickness of about 1.6 μm required for this work, a conventional spinner (S 1813 from Shipley) was used at 4000 rpm (acceleration 450 rpm s^{-1}) for a duration of 30 s.

Figure 4.27 The mask patterns used for the lithography process. While the bottom left pattern is the one that was used for the bottom part (supports), the other three patterns were used to fabricate top parts with 1, 3, and 4 hanging beams, respectively

Soft Baking
Immediately after the samples were spin coated, they were placed on a hotplate using a vacuum line to ensure hard contact, after which the temperature was set to 115 °C and the samples baked for 1 minute.

Exposure
To transfer the mask pattern to the photo resist, a conventional mask aligner (Karl Suss MA6) was used. The exposure time was selected to be 5 s, while a 270 W mercury lamp (in soft contact mode) was used.

Developing
For the developing step, the samples were agitated in a solution of photo resist developer (MF-319 from Shipley) for about 1 minute at room temperature. After developing, samples were inspected under a microscope to detect any probable defects. A picture of some of the samples after this developing step is shown in Figure 4.28.

Hard Baking
In order to reinforce the photoresist for subsequent etching processes, a hard baking step was required, in which samples were heated in an oven for 30 minutes at a temperature of 105 °C.

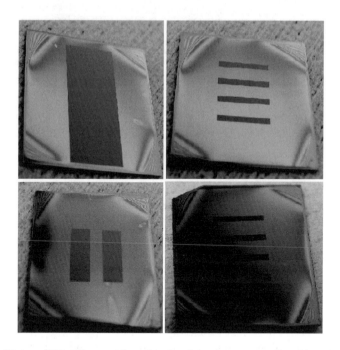

Figure 4.28   The samples after development process (See Plate 7)

4.2.3.6   Oxide Layer Etching

The conventional etchant for SiO_2 is BOE and comprises a mixture of HF (one part by volume)[2] and ammonium fluoride, NH_4F (six or seven parts by volume). With an etch rate of approximately 1000 min^{-1} at room temperature, BOE is a selective etchant for oxide. It does not etch silicon, so the removal of an oxide layer from a silicon crystal is self-limiting. Nevertheless, HF attacks photoresist to some extent, although the addition of ammonium fluoride reduces this effect. When end-point detection is required over silicon, we can make use of the wetting properties of silicon and of oxide. Oxide is hydrophilic and easily absorbs water. Silicon, on the other hand, is hydrophobic and repels water. Therefore, a completely etched silicon substrate, dipped in water, will shed the water instantly when removed. By contrast, a substrate with even a very thin layer of oxide on the surface will remain wet. In order to protect the back side of the samples, a special tape was used. Using a solution of BOE (7:1), 2 min was enough to etch the oxide layer. After rinsing in DI water, another inspection by microscope was undertaken. If remnants of SiO_2 were observed, the BOE etching could be repeated. The photoresist layer, as well as the protective tapes, was then removed in acetone.

4.2.3.7   Tetramethylammonium Hydroxide, TMAH Etching

One of the conventional anisotropic wet etchants for silicon is (TMAH). The solution used for this work was TMAH 25% from Moses Lake Industries. As shown in Figure 4.29, a

---

[2] HF with pH = 1 is extremely dangerous and must be treated cautiously. Before working with the chemicals, reading the associated MSDS is highly recommended.

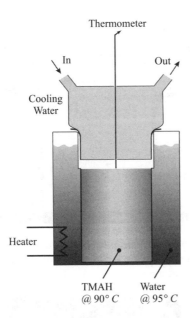

Figure 4.29   The TMAH etching setup used for micromachining

water bath (from VWR Scientific, Model 1235 PC) was used to control the temperature of the TMAH. To maintain the TMAH at 90 °C, the temperature of the water was set to 95 °C. A condenser at the top of the container prevented evaporation of TMAH. The condenser was kept cool by circulating the water at a flow rate of about 1 L min^{-1}. The samples were placed in a sample holder and inserted into the container. The previous characterization tests showed that etch rates of silicon and thermal oxide are equal to 0.670 μm min^{-1} and 1.15 A$^\circ$ min^{-1}, respectively. Therefore, to etch 180 μm of silicon (<100> direction), 4.5 hours were required. As previously mentioned, the sensor consists of two parts. One part, which served as the support, was etched for 1.5 hours to obtain an etch depth of 60 μm, while the top parts were maintained in TMAH for 4.5 hours to etch through and release the hanging beams. Samples of etched parts, in which Figure 4.30a is the top part with three beams and Figure 4.30b is the support, are shown in Figure 4.30.

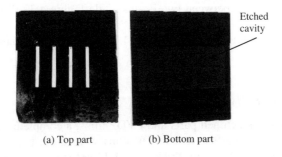

(a) Top part   (b) Bottom part

Figure 4.30   Samples of micromachined silicones

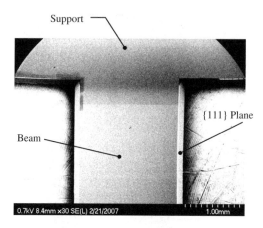

Figure 4.31 A micromachined beam and a support. (The (100) plane, which is parallel to the paper is shown. The {111} planes which make two sides of the beam are also illustrated)

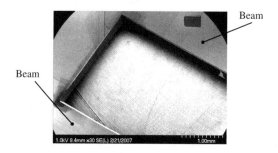

Figure 4.32 The void space between two adjacent beams

Due to the presence of silicon during TMAH etching, the existing oxide layer at the edges showed some irregularities. To achieve a uniform surface, it was decided to remove all existing oxide layers and repeat the oxidation step at the end. In this way a uniform and smooth surface of oxide layer was grown on the wafers.

To remove the oxide layer, the samples were treated as stated in Section 4.2.3.6, and then the process of wet oxidation, as explained in Section 4.2.3.4, was repeated. Figures 4.31 and 4.32 show an SEM (scanning electron micrograph) of the fabricated parts. Figure 4.31 shows the void between two beams while Figure 4.32 illustrates a beam and its support.

The {111} planes which are at an angle of 54.7° with the (100) plane can also be seen in Figure 4.31.

4.2.4 Sensor Assembly

Using the silicon micro-machined parts and PVDF films, a complete sensing module was assembled. As mentioned earlier, five PVDF films were used, two as supports and the other three for the beams. A 28 μm thick, uniaxial, and metalized PVDF film was used

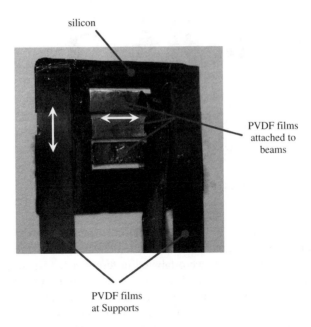

silicon

PVDF films
attached to
beams

PVDF films
at Supports

Figure 4.33 The top part of the sensor. PVDF films at the supports as well as films attached to the beams are shown. The direction of the PVDF films are also illustrated (See Plate 8)

from which two pieces of 22×4 mm film were cut in such a way that the drawn direction of each film was parallel to the film length. (Refer to Figure 4.17 in which arrows show the drawn direction.) This was done to minimize the output of the PVDF films at the supports due to any unwanted load and maximize the sensitivity of the output to the compressive load only. These two films were glued to the top of the supports. Of its total 22 mm length, 18 mm was used as the active area, and the remaining 4 mm used to connect the electrodes. Electrodes were cut from a thin copper foil and adhered to the PVDF films using conductive glue. For the suspended beams, three 2×12 mm PVDF films were used, in which each was cut in such a way that the drawn direction was in line with the PVDF film length. A picture from the backside of the top part is shown in Figure 4.33. After connecting electrodes to the films, they were glued to the top silicon part.

The wiring of these three films was passed through the empty space between the top and bottom parts. Finally, there were five pairs of wires, of which each pair belonged to one sensing element. The top part, upon which all the PVDF films were adhered, was glued to the bottom silicon part and then, for subsequent testing, the assembled sensor system was placed on a Plexiglas block. The hybrid assembled MEMS sensor is shown in Figure 4.34.

4.2.5 Testing and Validation: Softness Characterization

In order to validate the simulation results, a softness sensing experiment was conducted utilizing the microfabricated sensor. Prior to conducting the softness tests, it was required

Figure 4.34 The macromachined tactile sensor

to estimate the softness of test objects. In addition, it was also required to employ an additional quantification technique, rather than the conventional qualitative and heuristic approach, in which the test objects were labeled in terms very soft, soft, and so on. The durometer (Shore) softness test is one of the most commonly used hardness/softness tests for elastomeric materials. The durometer gauge measures the depth of surface penetration of an indenter of a given geometry. Since both force and deflection are measured, the durometer is typically used as a stiffness indicator of which a working concept is shown in Figure 4.35. When the durometer is pressed against an object, depending upon the softness, a calibrated spring contained within the indenter tip (whose shape and size can vary depending upon the type of durometer) is compressed and the softness number can then be read from a dial connected to the spring. The durometer hardness test is defined by ASTM D 2240, which covers seven different types of durometers: A, B, C, D, DO, O, and OO [19].

Table 4.2 summarizes the overall information about the A- and OO-type durometers that were used in this study. In this group, type OO was used for the measurement of very soft materials.

Figure 4.35 A picture of the durometer used in this study

Table 4.2   General specifications of durometer types A and OO

Scale of Durometer	Material to be used	Main spring force	Indenter shape
A	Soft elastometric materials, rubber, and rubber-like	822 gf (8.06 N)	Flat cone point, 35°
OO	Light foams, sponge rubber, and animal tissue	113 gf (1.11 N)	Sphere, 3/32″

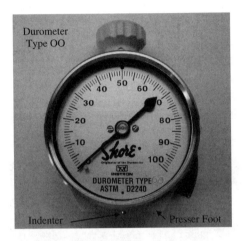

Figure 4.36   Durometer type OO

As shown in Figure 4.36, a commercially available durometer test consists of four components according to ASTM D 2240: the presser foot, indenter extension, indicating device (e.g., dial), and calibrated spring. The scale reading is proportional to the indenter movement. The soft material samples were measured using standard durometer types OO and A.

The successful application of the durometer in mechanical property evaluation of soft tissue such as skin is reported by Falanga and Bucalo [20]. The measurement of elasticity of tumors surrounded by soft tissue is reported in [21] and the range of elasticity of breast tumors was found to be between 150 and 990 kPa.

In addition to the softness measurements, the compressive Young's modulus of the objects for small strains (<10%) was measured using a conventional compression test. The relationship between the durometer dial reading and Young's modulus of the objects is shown in Figure 4.37. The Young's modulus of the harder object was measured as 6 MPa. The objects were then used to test the microfabricated sensor for softness sensing. Figure 4.38 shows the MEMS sensor during test with an electrodynamic shaker for applying dynamic loads. The soft objects were placed on the first beam. A compressive load was then applied to the top surface of the objects using a rigid plate at the end of a probe in order to mimic the grasping conditions.

The output of the PVDF films attached to all the three beams were recorded and compared. Four samples were tested. Three of them were considered as soft objects and

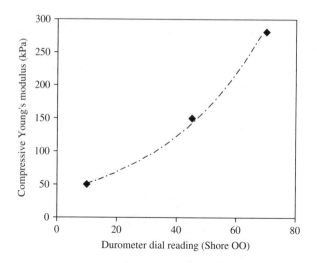

Figure 4.37   Durometer dial reading scale OO versus Young's modulus of soft objects is plotted

Figure 4.38   MEMS sensor being tested using an electrodynamic shaker

the last as a medium-hard object. In contrast to the finite element results, a moderate cross talk between three channels was observed. This cross talk can be attributed to both mechanical and electrical phenomena. The dynamic force applied to a beam can induce some vibration into adjacent beams, which causes some output from other channels.

On the electrical side, each sensing element comprised two electrodes and associated wiring, therefore 10 wires had to be used that could potentially affect other channels. Figure 4.39 shows the output of the middle PVDF films for three different soft objects with softness numbers 10, 43, and 70 on the Shore OO scale, when a distributed load of 111 kPa at 20 Hz was applied. The results of the finite element model, used to determine these material properties, are also added to the figure for comparison. As mentioned earlier, some level of cross talk was observed, recorded, and shown in Figure 4.40. Similar to the previous test, a sinusoidal distributed load at 20 Hz was applied to the first beam and

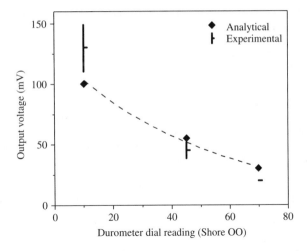

Figure 4.39   Plot of durometer dial reading scale OO versus Young's modulus of soft objects

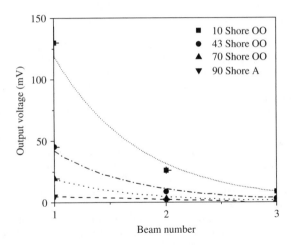

Figure 4.40   Plot of durometer dial reading scale OO versus Young's modulus of soft objects

this time the outputs of all beams were recorded. The results show that even when no load is applied to the second and third beams, an output voltage exists in the other two channels. The existence of a lump within the soft object would cause a point load.

References

1. Gray, B.L. and Fearing, R.S. (1996) A surface micromachined microtactile sensor array. Proceedings of the 1996 IEEE International Conference on Robotics and Automation, 1996, vol. 1, pp. 1–6.
2. Lazzarini, R., Magni, R., and Dario, P. (1995) A tactile array sensor layered in an artificial skin. Proceedings of the 1995 IEEE/RSJ International Conference on 1995 Intelligent Robots and Systems 95. 'Human Robot Interaction and Cooperative Robots', vol. 3, pp. 114–119.

3. Kolesar, E.S. and Dyson, C.S. (1995) Object imaging with a piezoelectric robotic tactile sensor. *Journal of Microelectromechanical Systems*, **4**, 87–96.
4. Metha, M. (1996) A micromachined capacitive pressure sensor for use in endoscopic surgery. Master of Applied Science, Engineering Science, Simon Fraser University.
5. Dargahi, J. and Payandeh, S. (1998) Surface texture measurements by combining signals from two sensing elements of a piezoelectric tactile sensor. Proceedings of the SPIE International Conference, 1998, pp. 58–62.
6. Fisher, H., Neisius, B., Trapp, R. *et al*. (1995) Tactile feedback for endoscopic surgery, *Interactive Technology and a New Paradigm for Health Care* (ed. K. Morgan, R.M. Satava, H.B. Seiburg, R. Mattheus, and J.P. Chris), IOC Press and Ohmsha, pp. 114–117.
7. Bicchi, A., Canepa, G., Rossi, D.D. *et al*. (1996) A sensorized minimally invasive surgery tool for detecting tissue elastic properties. Proceedings of the 1999 IEEE International Conference on Robotics and Automation, Minneapolis, MN, pp. 884–888.
8. Faraz, A., Payandeh, S., and Salvarinov, A. (1998) Design of haptic interface through stiffness modulation for endosurgery: theory and experiments. Proceedings of the 1998 IEEE International Conference on Robotics and Automation, 1998, vol. 2, pp. 1007–1012.
9. Howe, R.D., Peine, W.J., Kantarinis, D.A., and Son, J.S. (1995) Remote palpation technology. *Engineering in Medicine and Biology Magazine, IEEE*, **14**, 318–323.
10. Rebello, K.J. (2004) Applications of MEMS in surgery. *Proceedings of the IEEE*, **92**, 43–55.
11. Microline Surgical Company (2011) Grasper Tips, www.microlinesurgical.com, accessed 2011.
12. Hetenyi, M. (1964) *Beams on Elastic Foundation*, University of Michigan Press, pp. 38–49.
13. Ruska, W.S. (1987) *Microelectronic Processing*, McGraw-Hill.
14. Dargahi, J. (1993) *The Application of Polyvinylidene Fluoride as a Robotic Tactile Sensor*, Glasgow Caledonian University.
15. Dargahi, J. (1998) Piezoelectric and pyroelectric transient signal analysis for detection of the temperature of a contact object for robotic tactile sensing. *Sensors and Actuators A: Physical*, **71**, 89–97.
16. Dargahi, J. Eastwood, A.R., and Kemp, I.J. (1997) Combined force and position polyvinylidene fluoride (PVDF) robotic tactile sensing system. Proceedings of the SPIE International Conference on Sensor Fusion: Architectures, Algorithms and Applications, Orlando, FL, 20–25 April 1997, pp. 160–170.
17. Goodfellow, Technical Information for Polyvinylidenefluoride (PVDF), www.goodfellow.com, accessed 2007.
18. He, J.H., Sun, S., Ye, J., and Lim, T.M. (2006) Self -assembly carbon nanotubes on cantilever biosensor for sensitivity enhancement. *Journal of Physics: Conference Series*, **34**, 423–428.
19. Qi, H.J., Joyce, K., and Boyce, M.C. (2003) Durometer hardness and the stress–strain behavior of elastomeric materials. *Rubber Chemistry and Technology*, **76**, 419–435.
20. Falanga, V. and Bucalo, B. (1993) Use of a durometer to assess skin hardness. *Journal of the American Academy of Dermatology*, **29**, 47–51.
21. Ladeji-Osias, J.O. and Langrana, N.A. (2000) Analytical evaluation of tumors surrounded by soft tissue. Engineering in Medicine and Biology Society, 2000. Proceedings of the 22nd Annual International Conference of the IEEE, 2000, vol. 3, pp. 2114–2117.

5

Bulk Softness Measurement Using a Smart Endoscopic Grasper

5.1 Introduction

This chapter investigates the characteristics of the dynamic load that occurs when an object is grasped. In the realm of minimally invasive surgery (MIS) graspers, information about the amplitude, waveform, and frequency content of the load transferred to a soft object is important when performing experimental tests. In addition, the response of all smart endoscopic graspers (those equipped with tactile sensors) depends on the behavior and softness of the grasped object [1, 2].

A majority of studies on tactile applications are dedicated to force and pressure sensors [3, 4], although a couple of researchers have investigated and introduced tactile sensors capable of softness sensing, tailored for the measurement of local distribution of stress within the grasper [5, 6]. In Chapter 4, two endoscopic sensors were described that were capable of measuring distributed force or softness locally (along the grasper length) in order to detect foreign objects hidden in bulk soft tissue. These simple, smart sensorized graspers are also useful in obtaining information with regard to the interaction between soft objects and the sensor itself. The objective of this research is to obtain a profile of the repercussions of this phenomenon in finite element analysis, as well as the results of experiments using smart endoscopic graspers [7]. A simple, smart laparoscopic grasper is also presented and tested in order to determine the bulk softness of a grasped object by measuring the grasper jaw angle and the force applied.

5.2 Problem Definition

After an endoscopic tactile sensor is manufactured, we need to apply dynamic forces in an experimental test in order to characterize and calibrate the sensorized graspers. Although we have attempted to apply forces that are similar to those that grasper will actually encounter *in situ*, the initial problem will be to identify the constituent aspects of these force profiles. Because the amplitude and profile of the force depends, to some extent, on the geometry of the endoscopic grasper itself, we concentrated our attention on the influence that the object has on the grasper–object interaction in a typical MIS grasper. In

Tactile Sensing and Displays: Haptic Feedback for Minimally Invasive Surgery and Robotics, First Edition.
Saeed Sokhanvar, Javad Dargahi, Siamak Najarian, and Siamak Arbatani.
© 2013 John Wiley & Sons, Ltd. Published 2013 by John Wiley & Sons, Ltd.

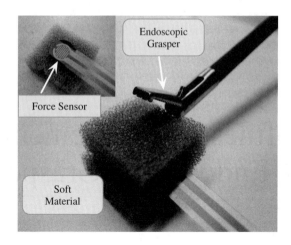

Figure 5.1   A typical grasper jaw, elastomeric material, and FSR sensor (See Plate 9)

the absence of soft tissue, we tested elastomeric materials which exhibit many of the same complexities and structures as are to be found in soft materials (e.g., material nonlinearity and viscoelasticity). A typical grasper jaw and soft object (i.e., elastomeric material) and force sensitive resistor (FSR) sensor are shown in Figure 5.1.

5.3   Method

As shown in Figure 5.1, in order to measure the force experienced by the grasped object, a force (piezoelectric) sensor was inserted inside the material to be grasped in order to measure the average force transmitted to the grasped object.This sensor is actually an FSR, from Interlink Electronics, whose resistance changes in inverse proportion to any force applied on its surface. The selected sensor was very small, thin, and flexible in order to be as unobtrusive as possible, and thereby avoid spurious readings.

For the experiments, as shown in Figure 5.2, an endoscopic grasper was equipped with two sensory devices. One was a potentiometer used to measure the grasper jaw angle. The second was an FSR affixed to the grasper handle in order to measure the force applied by the surgeon's thumb. The jaws clamp onto the material to be grasped, in which another FSR is inserted, in order to measure the force applied to the material.

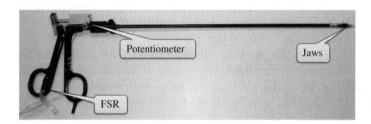

Figure 5.2   A picture taken from an endoscopic grasper equipped with two FSR sensors and a linear potentiometer (See Plate 10)

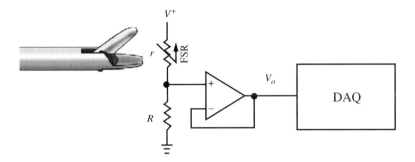

Figure 5.3 The electric circuit used for calibration and subsequent experiments

The signals from each force sensor (V_{o1} and V_{o2}), together with the signal from the potentiometer (V_p) and the supply voltage (V^+), were fed into a data acquisition card (DAQ).

To examine the linearity of the FSR sensor, and to obtain its calibration curve, an electric circuit, as shown in Figure 5.3, was used, in which the DAQ was used to analyze and record the measured data.

To experimentally determine the force–conductance relation for the FSR, several different standard weights were placed on the active area of the FSR and its conductance was measured. Experiments showed that the conductance was almost linear in relation to the applied force.

Using the parameters obtained from the electric circuit shown in Figure 5.3, the resistance of the FSR was found and its conductance C calculated from Equation 5.1 .

$$V_0 = \frac{R}{R+r} V^+ \Rightarrow C = \frac{1}{r} = \frac{1}{\left(\frac{V^+}{V_0} - 1\right) R} \tag{5.1}$$

where, r is the variable resistance of FSR (force-dependent), R is a known biasing resistor (constant); V^+ is the known voltage of an external power supply. The symbol V_o represents the FSR voltage as seen by DAQ. The signals from the FSR force sensors at the handle and tip of the grasper are denoted by V_{o1} and V_{o2}, respectively. At the calibration step, by applying different standard forces ranged from 1 to 9 N on the FSR, nine data points, as shown in Figure 5.4, were obtained. Using plotted force–conductance data points, the following expression (5.2) was obtained using the least squares regression method.

$$C = \frac{1}{r} = a.f + b \tag{5.2}$$

where $a = 0.079$ and $b = 0.016$.

Combining Equations 5.1 and 5.2, the following voltage–force relationship (Equation 5.3) was derived, from which the real applied force to grasped objects can be found:

$$f = \frac{1}{\left(\frac{V^+}{V_r} - 1\right) R.a} - \frac{b}{a} \tag{5.3}$$

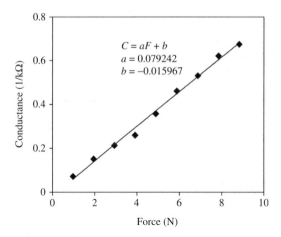

Figure 5.4 The conductance of the FSR versus applied force. A line was fitted to the measured data using the least squares fitting method

Figure 5.5 shows the block diagram of the program developed by LabVIEW whose processing software, using Equation 5.3, calculated the force applied to the grasped object and handle, as well as the jaw angle.

To investigate the profile of the load transferred to different materials, several soft materials were prepared and their softness measured using industrial durometers (for further information about durometers, refer to Chapter 4). The materials and their corresponding

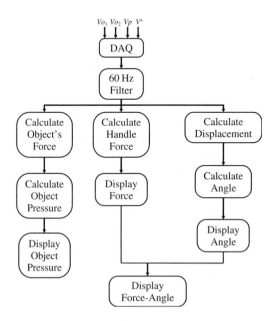

Figure 5.5 The block diagram of the program developed in LabVIEW

Table 5.1 The measured softness of five different materials used in the experiments

Material	Softness
ICF400	7 (Shore OO)
A2	10 (Shore OO)
EVA	45 (Shore OO)
B3	63 (Shore OO)
Silicone rubber-1 (White)	32 (Shore A)
Silicone rubber-2 (Blue)	52 (Shore A)
Silicone rubber-3 (Red)	71 (Shore A)
Silicone rubber-4 (Black)	89 (Shore A)

softness are presented in Table 5.1 . The first four materials (supplied by 3M) were considered to be very soft. These were tested using a Shore OO durometer, which is suitable for very soft materials, including soft tissue. The other four materials were different silicone rubbers selected from the standard type A (Shore A) test block kit (ASTM D2240, Instron Co.).

It should be noted that the materials were sorted from soft to hard (i.e., ICF 400 was the softest material in this group). The materials were then tested using the equipped grasper while an FSR sensor was inserted into the grasped object.

In conjunction with the force applied to the object and the dimensions of the force sensor itself, the average pressure was calculated. These parameters, of which a sample is shown in Figure 5.6, were then plotted in the time domain.

The processing software also calculated and plotted the relationship between the applied force and the jaw angle. Figure 5.7 shows the force–angle curve for an elastomer.

Two separate phases can be seen in this curve. In the first phase (loading), the open grasper jaws start closing and eventually apply full pressure to the material. The second phase (unloading) begins when the jaws start opening. In this phase the applied pressure is small. Therefore, most of the information on the material properties is gained during

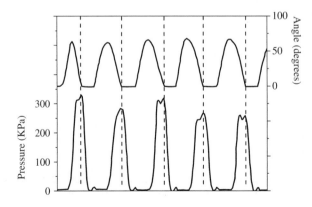

Figure 5.6 Samples of the recorded data. The angle of the grasper and registered pressure applied to the grasped soft object are shown. The pressure is expressed in kPa

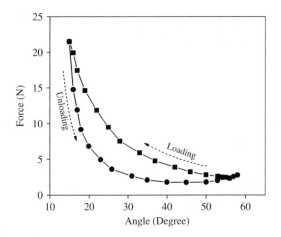

Figure 5.7   Force–angle curve for an elastomer (B3 $\frac{1}{2}$ inch)

the first phase. The form of this graph depends on the properties of the test material, from which valuable information can be derived. It can be shown that the speed of grasping can also affect the force–angle curve, although in this research work we attempted to ensure that a similar grasping speed was used for all test materials.

5.4   Energy and Steepness

In order to distinguish between the force–angle curves of different materials, average slope and energy were considered as the criteria. Figure 5.8 shows the experimental results for a silicone rubber.

The energy of the force–angle curve shown in Figure 5.8 is defined as the area below the loading curve from the beginning of the loading phase to the angle at which the energy

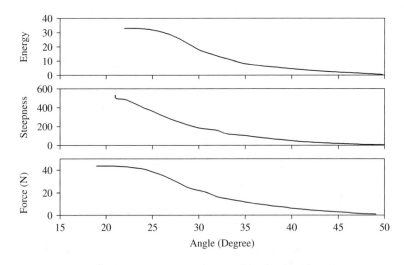

Figure 5.8   Force–angle curve for a silicone rubber, average steepness, and energy

is measured. The steepness is the slope of the force–angle curve averaged over the angle axis, again from the beginning of the loading phase to the measuring angle. For example, the steepness at 40° is the slope averaged from 50° (the beginning of the loading phase) to 40°. To compare the softness of the grasped objects, the energy, or steepness, of each object is compared at an arbitrary angle.

5.5   Calibrating the Grasper

The MIS grasper has, by necessity, to make several mechanical connections, of which all produce friction. The force used to overcome this friction can, itself, cause errors in determining the softness of the objects, especially when the objects themselves are probably very soft, since grasping soft objects requires only a small amount of force compared to the frictional force.

Figure 5.9 shows the experimental force–angle curve data for an empty grasper in both its loading and unloading phases. When an object was grasped, and in order to minimize the effect of friction, this extra force had to be subtracted from the real force that was applied to the object. In order to accomplish this, a formula was required to calculate the frictional force at each angle. Using the least squares error method, a curve fitted on the experimental data. Figure 5.10 shows the experimental data of several consecutive loading curves. The following fourth-order curve was fitted to the experimental data:

$$f = 6.9 \times 10^{-7}\theta^4 - 1.58 \times 10^{-4}\theta^3 + 1.31 \times 10^{-2}\theta^2 - 0.47\theta + 6.11 \qquad (5.4)$$

The result of subtracting the frictional force is shown in Figure 5.11. Figure 5.11a represents the grasping action for an empty grasper. Figure 5.11b shows the calculated force using Equation 5.4. Figure 5.11c shows the compensated curve when all frictional forces are completely removed.

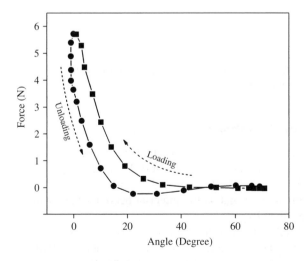

Figure 5.9   Experimental force–angle curve data for an empty grasper

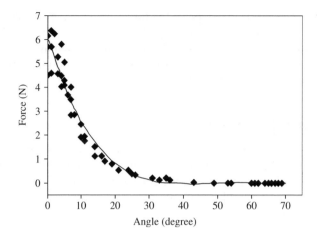

Figure 5.10   Experimental data of several consecutive loading curves and the fitted curve

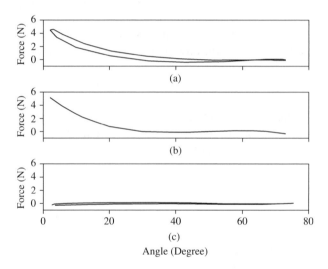

Figure 5.11   Subtracting the friction: (a) grasping action for an empty grasper; (b) calculated force using curve-fitting formula; (c) the compensated curve

5.6   Results and Discussion

The registered pressures which are transmitted to different soft materials are shown in Figure 5.12. As previously mentioned, ICF400 (Figure 5.12a) was very soft, and silicone rubber-2 (blue, Shore A = 52) (Figure 5.12e) was the hardest in this group. It can be concluded from Figure 5.12 that as the material becomes harder, the transferred pressure to the material becomes larger.

For example, in ICF400 (with a softness of 7 on the Shore OO scale) the maximum transferred pressure was 200 kPa, and for silicone rubber-2 (with a softness of 60 on Shore A) the maximum pressure was more than 600 kPa.

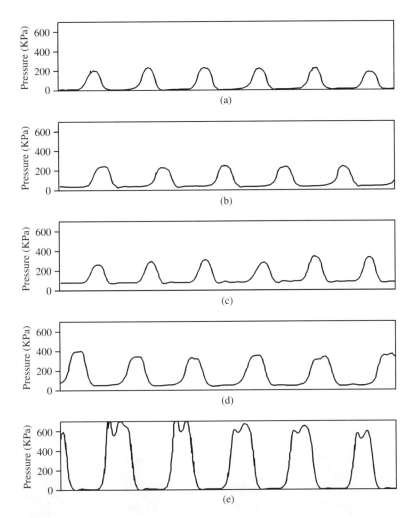

Figure 5.12   The force profile for five different soft materials: (a) ICF400, (b) A2FR, (c) H1N, (d) Silicone rubber-1, and (e) Silicone rubber-2

Another important conclusion drawn from Figure 5.12 is that the waveform of the transferred pressure also depends on the softness of the grasped material. For soft material, the pressure has a half-sinusoidal form. For harder materials, the pressure appears to be more like a pulse wave.

These force profiles can be associated with the response of soft materials grasping at low frequencies. When a soft material is grasped, it gradually becomes compressed and the resistance against this increases correspondingly. When the grasper starts to open, due to the same spring action, the soft material starts to expand so that it maintains its contact with the grasper jaws. This causes a smooth rise and fall in the force profile, which is similar to a sinusoidal wave. In fact, the frequency content of the force profile also shows that this force profile can be approximated to a sinusoidal wave. On the other hand, when grasping hard material, once the grasper touches the object the force

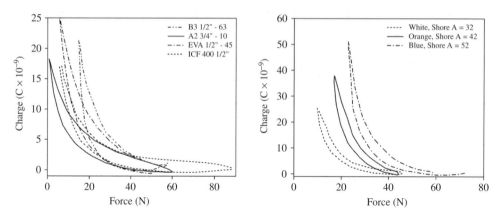

Figure 5.13 The force–angle curve for (a) ICF400, EVA, B3, and A2 and (b) silicone rubbers (white, Shore A=32; orange, Shore A=42; blue, Shore A=52) (See Plate 11)

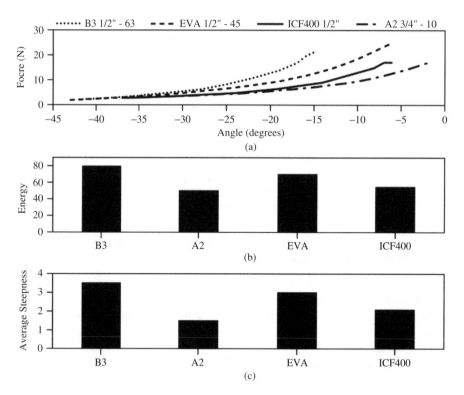

Figure 5.14 (a) The force–angle curve, energy, and steepness measured at 35°, (b) energy and (c) steepness

increases rapidly without significantly changing the grasper angle. In the opening stages, due to the rigidity of the grasped object, the grasper becomes quickly detached from the object, causing a force profile similar to a pulse wave. This conclusion is particularly important for experimental tests on smart graspers. This result confirms that the selection of a sinusoidal waveform to test the fabricated MIS tactile sensor is the best alternative in a laboratory test setup. It has always been argued that when a surgeon applies force to the grasper, no sinusoidal wave is produced, so all experiments using this waveform would be questionable. But the latter conclusion shows that for soft material, the load transferred to the grasped object is almost sinusoidal.

Figure 5.13 demonstrates the force–angle relationships for different materials. In these graphs, the force represents the load that is applied to the handle of the grasper and the angle is that of the grasper jaws. From Figure 5.13, it can be seen that in the loading phase of the curves, the average slope is proportional to the softness of the grasped material.

For softer material, the average slope is smaller than that of harder material. Figure 5.14 shows the energy and steepness for four elastomers. Both energy ($E = \sum_{\theta_1}^{\theta_2} F(\theta) . \Delta\theta$) and steepness ($S = \frac{1}{N} \sum_{\theta_1}^{\theta_2} \frac{F(\theta + \Delta\theta) - F(\theta)}{\Delta\theta}$) criteria plotted in Figure 5.14b,c can easily be used to arrange the softness of these materials. It is evident that EVA, H1N, A2FR, and ICF400 are arranged from hard to soft. Figure 5.15 shows the force–angle curves for the same materials, but the angle at which the energy and steepness are calculated is 25°.

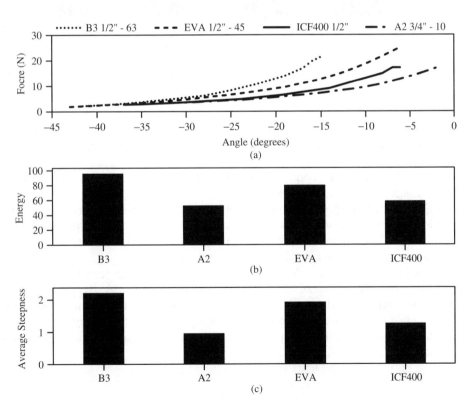

Figure 5.15 (a) The force–angle curve, energy, and steepness measured at 25°, (b) energy, and (c) steepness

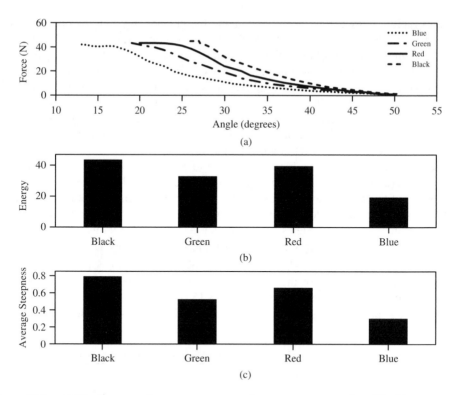

Figure 5.16 (a) The force–angle curve, energy, and steepness measured at 40°, (b) energy, and (c) steepness

This shows that the result of applying the criteria is independent of the angle to which the criteria are applied. Finally Figure 5.16 shows the force–angle curves of four silicone rubber samples.

Figure 5.16b,c shows the energy and steepness of the curves; it is still possible to use them to arrange the softness of the silicones.

As mentioned, the proposed system comprises an endoscopic grasper equipped with force sensors, a potentiometer for measuring the jaw angle, and a signal processing unit. It is shown that this system can, potentially, be used for softness characterization of the grasped tissue in laparoscopic surgery.

The experiment results show that for the soft materials, the transferred force from the grasper to the material is sinusoidal. As the material becomes harder, the pressure waveform changes from sinusoidal to pulse. Also it can be concluded that for harder materials, the load applied to the object becomes larger. Finally, using the force–angle graphs plotted for the test materials, it is shown that the grasped materials can be discriminated from each other based on their softness.

References

1. Sokhanvar, S., Ramezanifard, M., Dargahi, J. *et al*. (2008) The force and softness signature in minimally invasive surgery graspers, CSME/SCGM forum, Ottawa, Canada.
2. Ramezanifard, M., Sokhanvar, S., Xie, W. *et al*. (2008) Smart endoscopic tool for the measurement of force and softness of grasped object in minimally invasive surgery. *ASME Conference Proceedings*, 2008, 25–26.
3. Gray, B.L. and Fearing, R.S. (1996) A surface micromachined microtactile sensor array, Proceedings of 1996 IEEE International Conference on Robotics and Automation, vol. 1, pp. 1–6.
4. Wang, L. and Beebe, D.J. (2000) A silicon-based shear force sensor: development and characterization. *Sensors and Actuators A: Physical*, **84**, 33–44.
5. Omata, S., Murayama, Y., and Constantinou, C.E. (2004) Real time robotic tactile sensor system for the determination of the physical properties of biomaterials. *Sensors and Actuators A: Physical*, **112**, 278–285.
6. Shikida, M., Shimizu, T., Sato, K., and Itoigawa, K. (2003) Active tactile sensor for detecting contact force and hardness of an object. *Sensors and Actuators A: Physical*, **103**, 213–218.
7. Ramezanifard, M. (2008) *Design and Development of New Tactile Softness Displays for Minimally Invasive Surgery*, M.A.Sc. MR40921, Mechanical and Industrial Engineering Department, Concordia University (Canada), Canada.

6

Lump Detection

6.1 Introduction

Palpation is the act of feeling with the hand. Primarily, it is the application of the fingers with light pressure to the surface of the body in order to determine the condition of the parts beneath for the purposes of physical diagnosis. It is this crucial and fundamental feature, however, that is missing in the current field of haptics as well as MIS (minimally invasive surgery) and robot-assisted surgery. Palpation is routinely used by surgeons in open surgery to differentiate between abnormal and normal tissues, or to detect the type of tumor based on biological tissue composition and consistency, which often varies from one tissue to another due to various diseases [1]. The localization of hidden anatomical features has been the subject of some research [2–6] though these studies have largely been focused on breast cancers, hence these findings cannot be directly used in either haptics or robotic surgical applications. Kattavenos *et al.* [7] reported the development of a tactile sensor for recording data when the sensor is swept over a phantom sample containing simulated tumors although no information regarding the size and depth of any lump was extracted from this data.

This chapter presents a hyperelastic finite element model of a slab of soft tissue embedded with a lump, from which the effects of various parameters, such as tissue thickness, size of the lump, relative Young's modulus of tissue and lump, and the distance of the lump from the surface of the tissue is investigated. These results will help us to better interpret any tactile information that is collected from a smart endoscopic grasper holding tissue in which a lump is embedded.

6.2 Constitutive Equations for Hyperelasticity

The conventional theory of elasticity is primarily based on infinitesimal strains, in which linear elastic assumptions are applied. For finite deformations, however, these assumptions are not valid and, in general, the response of the material is different from that of the linear theories. Any material that can experience a recoverable large elastic strain is referred to as being hyperelastic and for which the following Helmholtz free-energy function:

$$\psi = \psi(C) = \psi_I [I_1(C), I_2(C), I_3(C)] = \psi_\lambda(\lambda_1, \lambda_2, \lambda_3)$$

Tactile Sensing and Displays: Haptic Feedback for Minimally Invasive Surgery and Robotics, First Edition.
Saeed Sokhanvar, Javad Dargahi, Siamak Najarian, and Siamak Arbatani.
© 2013 John Wiley & Sons, Ltd. Published 2013 by John Wiley & Sons, Ltd.

can be applied in such a way that:

$$S_{ij} = 2\frac{\partial \psi}{\partial C_{ij}} \qquad (6.1)$$

where I_1, I_2, and I_3 are the strain invariants of the symmetric right Cauchy–Green tensor C, and λ_a, $a = 1,2,3$ are the principal stretches. The Second Piola-Kirchoff stress tensor S, is related to the Cauchy stress tensor σ, by $\sigma = J^{-1}FSF$, where F is the deformation gradient of the motion, that is, $F(X,t) = \text{grad } x$. It can be shown that the relationship between the Cauchy–Green tensor C and the deformation gradient F is: $C = F^T F$. The Jacobian of the motion, J, is defined as $J = \det[F_{ij}]$.

In this chapter, to denote scalar, vector, and tensor quantities we use uppercase letters when they are evaluated in the reference configuration, and lowercase letters for corresponding quantities in the current configurations. For example, symbols X and x represent the positions in the referential (original) and deformed (current) configurations, respectively.

The general form of the Cauchy stress tensor σ for an isotropic and incompressible material can be derived as [8]:

$$\sigma = -pI + 2\frac{\partial \psi}{\partial I_1}C - 2\frac{\partial \psi}{\partial I_2}C^{-1} \qquad (6.2)$$

where I_1, I_2 (and I_3), the invariants of C_{ij} are:

$$\begin{cases} I_1 = \lambda_1^2 + \lambda_2^2 + \lambda_3^2 \\ I_2 = \lambda_1^2\lambda_2^2 + \lambda_2^2\lambda_3^2 + \lambda_3^2\lambda_1^2 \\ I_3 = \lambda_1^2\lambda_2^2\lambda_3^2 = J^2 \end{cases} \qquad (6.3)$$

The symbol J is also a measure of compressibility, that is, the ratio of the deformed elastic volume to the undeformed (reference) volume of material. For incompressible material, therefore, a kinematic constraint, namely $J = \lambda_1\lambda_2\lambda_3 = 1$ can be considered. Hence, as can be seen in Equation 6.2, the two principal invariants I_1 and I_2 are the only independent deformation variables.

6.2.1 Hyperelastic Relationships in Uniaxial Loading

All strain energy density functions that are introduced for hyperelastic materials contain some unknowns referred to as *material constants* and it is important these are accurately computed for the material under the test conditions. The conventional way of deriving these constants is by using the experimental stress–strain data. Technically, it is recommended that these test data should be taken from several modes of deformation over a wide range of strain values. The number of modes of deformation in experimental tests should be at least as many deformation states as will be experienced during the analysis [9].

For the present study, the test data of compression, which is the dominant mode in MIS grasping, was used in order to find the material constants of the strain energy function. For a uniaxial compression, $\lambda_1 = \lambda_{\text{applied}} = \lambda$ is the stretch in the direction being loaded.

For an incompressible isotropic material in an unconstrained compression test, the stretch in the other two directions are equal, $\lambda_2 = \lambda_3$ and also $\sigma_2 = \sigma_3 = 0$.

From the introduced relationships among the principal stretches and the compressibility condition ($J = \lambda_1 \lambda_2 \lambda_3 = 1$) we arrive at:

$$\lambda_2 = \lambda_3 = \lambda^{-\frac{1}{2}} \tag{6.4}$$

For the uniaxial loading the motion χ can be expressed by the explicit equations:

$$\begin{cases} x_1 = \lambda X_1 \\ x_2 = 1/\sqrt{\lambda} X_2 \\ x_3 = 1/\sqrt{\lambda} X_3 \end{cases} \tag{6.5}$$

Therefore, the deformation gradient, $F_{aA} = \frac{\partial \chi_a}{\partial X_A}; a, A = 1, 2, 3$ has the form:

$$[F] = \begin{bmatrix} \lambda & 0 & 0 \\ 0 & \sqrt{\lambda} & 0 \\ 0 & 0 & \sqrt{\lambda} \end{bmatrix} \tag{6.6}$$

The Right Cauchy–Green tensor, C, can be obtained from Equation 6.6 as follows:

$$[C] = [F]^T [F] = \begin{bmatrix} \lambda^2 & 0 & 0 \\ 0 & \lambda^{-1} & 0 \\ 0 & 0 & \lambda^{-1} \end{bmatrix} \tag{6.7}$$

The strain invariants, therefore, can be calculated directly from the principal stretches given by Equation 6.3 or from the Cauchy–Green tensor:

$$\begin{cases} I_1 = tr(C) = \lambda^2 + 2\lambda^{-1} \\ I_2 = \frac{1}{2} \left\{ [tr(C)]^2 - tr(C^2) \right\} = 2\lambda + \lambda^{-2} \\ I_3 = \det(C) = 1 \end{cases} \tag{6.8}$$

Considering relations given in Equation 6.7, the Cauchy stress given by Equation 6.2 will be reduced to the principal stresses:

$$\sigma_{11} = -p + 2 \frac{\partial \psi}{\partial I_1} \lambda^2 - 2 \frac{\partial \psi}{\partial I_2} \lambda^{-2} \tag{6.9}$$

$$\sigma_{22} = -p + 2 \frac{\partial \psi}{\partial I_1} \lambda^{-1} - 2 \frac{\partial \psi}{\partial I_2} \lambda = 0 \tag{6.10}$$

Subtracting Equation 6.10 from Equation 6.9, we find that:

$$\sigma_{11} = 2 \left(\lambda^2 - \lambda^{-1} \right) \left[\frac{\partial \psi}{\partial I_1} + \frac{\partial \psi}{\partial I_2} \lambda^{-1} \right] \tag{6.11}$$

Therefore, by having a suitable strain energy function, the derivation of ψ with respect to the invariants can be obtained and the stresses (here σ_{11}) can be found. For example, consider a strain energy function of the following general form [10]:

$$\psi = \sum_{\substack{i=0 \\ j=0}}^{\infty} C_{ij} \left(I_1 - 3 \right)^i \left(I_2 - 3 \right)^j \tag{6.12}$$

which can be used for a wide range of applications. For a stress-free condition in the reference configuration (undeformed state), ψ should be equal to zero. Since in the reference configuration, $\lambda_1 = \lambda_2 = \lambda_2 = 1$, and from Equation 6.8 , $I_1 = I_2 = 3$, therefore, I_1 and I_2 are subtracted by 3. In addition, for the same reason, the constant C_{00} (when $i = j = 0$) must be zero.

In the above equation when $i = 0, j = 1$ and $i = 1, j = 0$, the two-term Mooney–Rivlin equation is obtained, that is, $\psi = C_{10}(I_1 - 3) + C_{01}(I_2 - 3)$. By substitution of partial derivation of ψ with respect to I_1, that is, $\partial\psi/\partial I_1 = C_{10}$ and I_2, that is, $\partial\psi/\partial I_2 = C_{01}$ into Equation 6.11, we find that:

$$\sigma_{11} = 2 \left(\lambda^2 - \lambda^{-1} \right) \left(C_{10} + C_{01}\lambda^{-1} \right) \tag{6.13}$$

The constants C_{01} and C_{10} in Equation 6.13 can be obtained from the stress–strain relationship of the material. To develop a more accurate model, more terms can be used. For instance, the three-term Mooney–Rivlin model for incompressible materials, which was used for the present study, is in the form:

$$\psi = C_{10} \left(I_1 - 3 \right) + C_{01} \left(I_2 - 3 \right) + C_{11} \left(I_1 - 3 \right) \left(I_2 - 3 \right) \tag{6.14}$$

In the uniaxial stress–strain compression test, a cubic specimen of elastomeric material with dimensions $17 \times 17 \times 19$ mm was cut out of a sheet and placed between two plates designed for compression tests. Then, with a displacement rate of 2 mm s^{-1}, the sample was compressed until the force reached to the machine's force limit (40 N) and the test was terminated. The compression tests were undertaken using an MCR (modular compact rheometer) 500 from Physica, Anton Paar. In order to reach to a stable and repeatable condition, before recording the test results, three compression tests were done. The output results of the corresponding software were in the form of force and displacement, which were converted to the engineering strain and stress using the area and the initial thickness of the specimens. The dependency of the results on the strain rate was examined by repeating the test at different strain rates. For the selected rubber-like material, negligible difference between results was observed. Three samples were tested under the above-mentioned conditions and the results showed that the maximum error was less than 4% among the samples. The numerical values were averaged and the result was plotted, as shown in Figure 6.1. The shape of the curve is similar to those of compression tests done on abdominal organs [1], which illustrates the similarity of the hyperelastic behavior of elastomeric material to that of real tissue.

In order to find the constants of the model, using the least squares fitting procedure, the averaged curve was used. The experimental values of engineering stress–strain were then compared with the curve obtained from the optimization procedure for the three-term Mooney–Rivlin model, as illustrated in Figure 6.1.

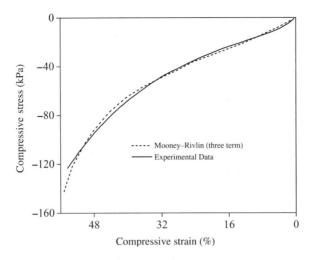

Figure 6.1 Comparison between experimental results and three-term Mooney–Rivlin strain energy function. The horizontal and vertical axes are engineering strain and engineering stress, respectively

Table 6.1 The constants of two- and three-term Mooney–Rivlin Models (kPa)

Model	N	C_{10}	C_{01}	C_{11}	Residual
Mooney–Rivlin	2	31.33	−8.92	–	7.34
Mooney–Rivlin	3	68.49	−39.88	5.71	4.7

It should be noted that initially a two-term model was tried. Nevertheless, the three-term model was preferred due to its better approximation. The numerical values of the constants for the two- and three-term Mooney–Rivlin models are summarized in Table 6.1.

6.3 Finite Element Modeling

A nonlinear three-dimensional finite element model using ANSYS commercial software was developed. The 'mixed u-p formulation' was used to solve the problem because of its superior results in solving hyperelastic incompressible problems compared with the 'pure displacement formulation' [12]. Although the displacement formulation is computationally efficient, its accuracy is dependent on Poisson's ratio or bulk modulus. In this formulation, volumetric strain is determined from derivatives of displacements, which are not as accurately predicted as the displacements themselves. For nearly incompressible materials, in which Poisson's ratio is close to 0.5 or the bulk modulus approaches infinity, any small error in the predicted volumetric strain will appear as a large error in the hydrostatic pressure and subsequently in the stresses. This error, in turn, will also affect the displacement prediction since external loads are balanced by the stresses. This may result in smaller displacements than expected for a given mesh (which is called locking) or, in some cases, result in no convergence at all. This problem is also encountered in fully incompressible deformation, such as for fully incompressible hyperelastic materials. In the mixed u-p formulation, in which the hydrostatic pressure $\bar{P}$ is interpolated on the element level, and solved on the global level independently in the same way as displacement, these difficulties are eliminated. In this method, the contribution of hydrostatic pressure, referred to as 'volumetric response,' is separated from the 'deviatoric response.' Therefore, the stress, for instance has to be updated by:

$$\sigma'_{ij} = \sigma_{ij} - \delta_{ij}\bar{P} \qquad (6.15)$$

where the prime indicates the deviatoric component of the Cauchy stress tensor. Alternatively, the deviatoric component of the deformation tensor e_{ij} can be expressed as:

$$\varepsilon'_{ij} = \varepsilon_{ij} - \frac{1}{3}\delta_{ij}e_v \qquad (6.16)$$

where $e_v = \delta_{ij}e_{ij} = e_{ii}$ and:

$$e_{ij} = \frac{1}{2}\left(\frac{\partial u_i}{\partial x_j} + \frac{\partial u_j}{\partial x_i}\right) \qquad (6.17)$$

in which u_i and x_i are the displacement and coordinate in the current configuration, respectively. In addition, for a fully incompressible hyperelastic material, the volume constraint is the incompressible condition [13–15]:

$$1 - J = 0 \qquad (6.18)$$

where, $J = |F_{ij}| = dv/dV_0$, in which $|F_{ij}|$ is the determinant of deformation gradient tensor and V_0 is the original volume. The constraint Equation 6.18 is introduced to the internal virtual work by the Lagrangian multiplier $\bar{P}$. Finally, differentiating the augmented internal virtual work gives the stiffness matrix in the form:

$$\begin{bmatrix} K_{uu} & K_{up} \\ K_{pu} & K_{pp} \end{bmatrix}\begin{Bmatrix} \Delta u \\ \Delta\bar{P} \end{Bmatrix} = \begin{Bmatrix} R \\ 0 \end{Bmatrix} \qquad (6.19)$$

where Δu and $\Delta\bar{P}$ are displacement and hydrostatic pressure increments, respectively.

The stiffness submatrices can be obtained from the following equations [16]:

$$K_{uu} = \int_V [B_D]^T [C'][B_D]\,dV \qquad (6.20)$$

$$K_{up} = K_{up}^T = -\int_V [B_V]^T [N_p]\,dV \qquad (6.21)$$

$$K_{pp} = -\int_V [N_p]^T \frac{1}{k}[N_p]\,dV \qquad (6.22)$$

where $[C']$ is the stress-strain matrix for the deviatoric stress and strain components. The matrix $[N_p]$ can be obtained from the additional interpolation introduced for hydrostatic pressure:

$$p = [N_p]\{\bar{P}\}.$$

The deviatoric strain-displacement matrix $[B_D]$ is given by:

$$\{\varepsilon'\} = [B_D]\{d\}$$

in which $\{d\}$ is the vector of nodal displacements. In element formulation, material constitutive law has to be used to create the relation between stress increment and strain increment. The constitutive only reflects the stress increment due to straining. In this case, however, the Cauchy stress cannot be used because it is affected by the rigid body rotation and is not objective (not frame invariant). Therefore, an objective stress is needed in order that it can be applied in constitutive law. The Jaumann–Zaremba rate of the Cauchy stress, expressed by McMeeking and Rice [17], is one of them and is defined as follows [18]:

$$\dot{\sigma}_{ij}^J = \dot{\sigma}_{ij} - \sigma_{ik}\dot{\omega}_{jk} - \sigma_{jk}\dot{\omega}_{ik} \tag{6.23}$$

where $\dot{\sigma}_{ij}^J$ is the Jaumann–Zaremba rate of the Cauchy stress, $\dot{\omega} = \frac{1}{2}\left(\frac{\partial v_i}{\partial x_j} - \frac{\partial v_j}{\partial x_i}\right)$ is the spin tensor and $\dot{\sigma}_{ij}$ is the time rate of the Cauchy stress. Alternatively, using the constitutive law, the stress change due to straining can be expressed as:

$$\dot{\sigma}_{ij}^J = C_{ijkl}d_{kl} \tag{6.24}$$

where C_{ijkl} is the material constitutive tensor and d_{kl} is the rate of deformation tensor, given by:

$$d_{ij} = \frac{1}{2}\left(\frac{\partial v_i}{\partial x_j} + \frac{\partial v_j}{\partial x_i}\right) = d_{ji} \tag{6.25}$$

in which v_i is the velocity. By substituting Equation 6.24 into Equation 6.23 , the Cauchy stress tensor can be obtained as:

$$\sigma_{ij} = C_{ijkl}d_{kl} + \sigma_{ik}\dot{\omega}_{jk} + \sigma_{jk}\dot{\omega}_{ik} \tag{6.26}$$

In the present study, in order to model the portion of tissue that is being held by an MIS grasper, a cube containing a stiffer object is considered for the geometry under study. The bulk soft tissue, as well as the tumor, was modeled using a 10-node tetrahedral Solid187 element which has a quadratic displacement behavior and is well suited to modeling irregular meshes. Each node has three DOF; translations in the nodal x, y, and z directions. This element has mixed formulation capability for simulating deformation of nearly incompressible elastoplastic materials, and fully incompressible hyperelastic materials. Although soft tissue is assumed to be hyperelastic, tumor is considered to be isotropic. The strain energy function (ψ) for soft material was selected to be the three-term Mooney–Rivlin model [19] due to its satisfactory performance in compression states of stress. The three coefficients of the model for incompressible isotropic hyperelastic materials were obtained from the implemented experimental stress–strain data using the least squares optimization method.

6.4 The Parametric Study

As mentioned in Chapter 5, a number of parameters have a significant effect on the output of sensors positioned on the contact surface. In general, some combinations of such parameters produce similar stress distributions on the contact surface. Therefore, some

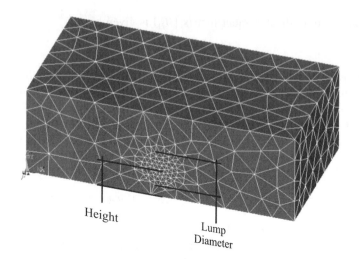

Figure 6.2   A half-model of soft tissue which contains a lump (See Plate 12)

restrictive assumptions and relationships are required in order to formulate the response of the system with respect to the variation of inputs. This is particularly important when constructing an inverse model in which information about a lump could be extracted from the tactile image or stress distribution on the contact surface. Therefore, introducing substitutive parameters obtained by combining several parameters is preferred.

As an example of parameter reduction: in general at least three parameters are required to determine the size of a lump, namely height, width, and depth. Since tumors can largely be approximated as spherical features, the number of parameters to characterize the size of the tumors or lumps can be reduced to one value, the lump radius. The cross section of the finite element model of a hidden mass embedded in soft tissue is shown in Figure 6.2.

In the following sections, the impact of variation of each involved parameter is presented. To demonstrate the variations of the tactile image in each case study, and to produce clearer results than 3D graphs, the stress distribution recorded on a path defined by a straight line passing from the middle of the bottom surface was recorded and plotted. The variation of pressure is represented by the pressure ratio P/P_{in}, in which P is the pressure distribution across the contact surface caused by the lump and P_{in} is the applied pressure. This ratio, therefore, defines how much the pressure distribution is influenced by the lump and other associated parameters.

6.4.1   The Effect of Lump Size

It is evident that early detection of any medical condition affords greater opportunities for healthcare workers to provide effective intervention. Therefore, in this regard, early detection of tumors (lumps) ranks as being among the most important since, particularly in such cases, early diagnosis is often crucial to eventual efficacious treatment and cure. In our simulation, spherical lumps of varying sizes were implanted in bulk soft tissue to show their effect on the pressure distribution and consequently on the system output.

In this set of simulations, the diameter of the lump was changed from 2 to 8 mm corresponding to $D/T \approx 0.2$ to $D/T \approx 0.8$ in which D and T are the diameter of the lump

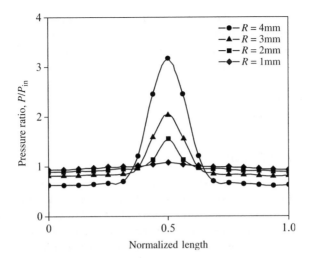

Figure 6.3 The pressure distribution at the contact surface when the lump radius was increased from $R = 1$ to 4 mm and $P_{in} = 5$ kPa

and thickness of tissue, respectively. For each ratio, the pressure distribution at the contact surface was calculated and is shown in Figure 6.3. In these simulations, the load was applied in the form of displacement (strain). The surface nodes were coupled together and a known displacement (e.g., $U_y = 1$ or 4 mm) was applied to the top surface of the tissue.

The center of the lump was considered to be in the middle of the tissue. From the demonstrated results, shown in Figure 6.4, it can be deduced that with the increase of lump size, the maximum pressure value increases almost linearly for relatively small and

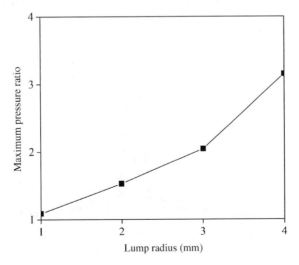

Figure 6.4 The pressure distribution at the contact surface when the lump radius was increased from $R = 1$ to 4 mm

medium-sized lumps, whereas the peak pressure elevates nonlinearly for larger lumps. In Figure 6.3, the curve corresponding to a tumor size of $R = 1$ mm, with a low peak value, indicates that small-sized lumps are difficult to detect. As shown in the next simulation, the maximum pressure ratio is also highly sensitive to the depth of the hidden lump. A lump with a large diameter exhibits two effects: it covers a larger part of the sensing area and also its outer surface is closer to the sensing area. The effect of lump depth is discussed in the next section.

6.4.2 The Effect of Depth

The depth of the lump, H, is the distance of the lump from the bottom contact (sensitive) surface and plays an important role in the amplitude, as well as the shape, of the pressure contour. Figure 6.5 shows how the stress distribution at the contact surface varies with the depth of the lump. In this test, the diameter of the lump was kept constant, that is, $D = 4$ mm, while the lump height was changed from 2 to 8 mm from the contact surface, in 2 mm steps, and a 10% compressive strain ($P_{in} \approx 10$ kPa) was applied through the top surface.

As can be seen in Figure 6.5, the maximum pressure drops dramatically when the lump distance increases. This trend is shown in Figure 6.6, in which the peaks of each pressure distribution are plotted against the depth of lump. For cases in which the distance is 6 and 8 mm, the peak pressures are almost identical. However, for lumps farther away, the pressure curve covers a larger area. For a lump at height $H = 2$ mm a slight undershoot was observed, which could be a distinctive sign for marginal lumps. By using this information in the design stage, it is possible to calculate the minimum spatial resolution for sensing elements, in order to have a good approximation of lump depth. The difficulty in detecting deep lumps suggests that several attempts, from different orientations, might well be required in order to localize the lump. Furthermore, it can be shown that smart graspers with two sensorized surfaces perform better when it comes to detecting marginal lumps.

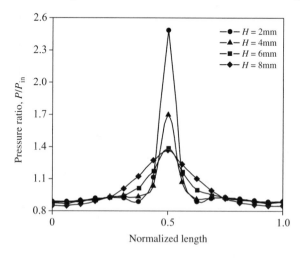

Figure 6.5 The effect of lump depth on the stress distribution when $R = 2$ mm and the center of the lump is changed from 2 to 8 mm from the contact (bottom) surface

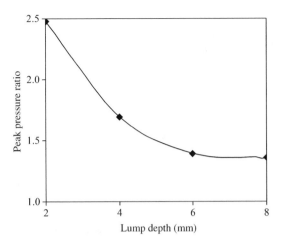

Figure 6.6 Peak pressures for different lump depths when, $R = 2$ mm and the center of the lump is changed from 2 to 8 mm from the contact surface

6.4.3 The Effect of Applied Load

The magnitude of the applied load also affects the pressure distribution. The results of the simulations when the applied load was changed from $U_y = 1$ to 4 mm are shown in Figure 6.7, from which it can be seen that the maximum pressure increases in proportion to the magnitude of the applied load. This graph also demonstrates that an increase in applied load does not affect significantly the area under the bell-shaped pressure contour. However, this dependency shows that for softness sensing or lump detection in addition to the stress profile, the magnitude of the total applied load must also be measured.

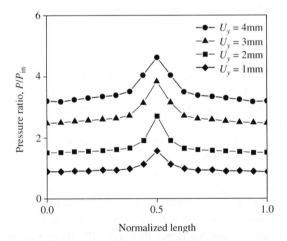

Figure 6.7 The variation of stress profile due to the variations of applied load. The values are normalized, based on the background pressure for $U = 1$ mm. In these simulations a lump with radius $R = 2$ mm was located in the middle of the tissue

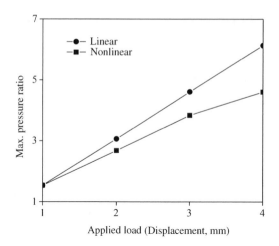

Figure 6.8 The maximum pressures obtained from nonlinear analysis against those of linear analysis. Simulation conditions are similar to that of Figure 6.7

To exemplify the difference between the results obtained from the nonlinear hyperelastic analysis and those of linearized relationships, we have compared the maximum pressure values computed for various applied displacements with those that are obtained from linear analysis.

This comparison, which is depicted in Figure 6.8, reveals that linear analysis shows higher values and, secondly, that this difference increases with higher deformation loads. This result is in agreement with the nonlinear behavior of the tested elastomeric material illustrated in Figure 6.1. This figure demonstrates that, for infinitesimal deformations, a straight line could be fitted to the experimental data, although, for larger deformations, this linear fitting can lead to significant errors.

6.4.4 The Effect of Lump Stiffness

The stiffness of the lump and tissue are also effective parameters in the studied case. Several researchers [3, 5] have used the ratio of Young's modulus of the lump to that of the soft tissue, (E_L/E_T), to analyze the relative variations of stiffness of lump with respect to that of the tissue.

This present analysis, considering nonlinear properties for the tissue, on the one hand, and assuming linear elastic isotropic property for the lump, on the other, makes the comparison between the stiffness of the lump and tissue complicated. Therefore, for this part of the analysis, the linear elastic modulus of the tissue was considered to be $E_T = 15$ kPa, close to the values reported for typical soft tissues [20]. Then, the Young's modulus of the lump, (E_L) was changed from 15 kPa to 15 MPa. The results, shown in Figure 6.9, indicate that for a Young's modulus larger than 1.5 MPa there was no discernible difference in stress profile. Indeed, the difference between the maximum pressures for $E_L = 150$ kPa and $E_L = 15$ MPa is less than 7%. It can be concluded that the system output is not very sensitive to E_L for $E_L > 10E_T$. This is useful in reducing the number of

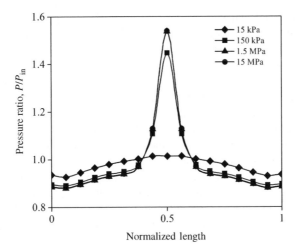

Figure 6.9   The stress distribution for different E_L, the Young's modulus of the lump while the Young's modulus of tissue was kept constant at E_T = 15 kPa. The applied load was U_y = 1 mm and the lump was considered to be in the middle of the tissue

unknowns for this problem. Since the reported Young's modulus for lumps, such as in breast cancer, is normally 10 times higher than that of the tissue [5], this variable can be regarded as being a constant.

6.5   Experimental Validation

Validation is an important step to ensure that the numerical analysis was performed appropriately and that correct results were obtained. Nevertheless, performing experiments is not always easy. For instance, preparing spherical lumps with exact dimensions on one hand, and carving out spherical voids from bulk soft tissue that have exactly the same dimensions is by no means a simple task. Any gaps between lumps and bulk tissue due to inaccurate carving of the tissue yield erroneous results and pressure distribution. It is also challenging to accurately adjust the center of the lump to the intended positions.

Creating identical boundary conditions in experiments and in the finite element model is another challenging task. In addition, frictional forces could play an important role in many experiments and are very difficult to measure and implement, both in the experiments and the finite element analysis. Dimensional and geometrical differences between the FEM and experiments are highly likely and could potentially be a source of error. The utilized sensing elements are also important to the final result. In this study, an array of the piezoelectric PVDF (polyvinylidene fluoride) sensing elements was used to register the stress distribution over the contact area. Therefore, extreme care was used to ensure that each prototype device contained an identical sensing element. However, after manufacturing, some discrepancies among the sensing elements were observed. Therefore, in the initial stage, the elastomer without any lump was placed on the sensing array and the outputs of the sensors were tested and calibrated. In this present study, due to the difficulties just mentioned, only two different combinations of parameters were tested.

The elastomeric material that was primarily used in the compression tests was also used in the experimental work as soft tissue. Two acrylic balls of different sizes (3 and 4 mm), simulating the lumps, were inserted into the hollow spaces that were carefully carved out of the bulk elastomeric. To change the depth of the lumps, several layers of the elastomeric material were cut into the same dimensions but different thicknesses. The lumps were placed in one of the layers so that other elastomeric layers could be used as spacers to increase or decrease the distance of the lump layer from the top or bottom surfaces, while the total thickness was kept constant. A dynamic load was applied by a shaker, which was driven by a power amplifier and a signal generator. To register the pressure distribution on the contact surface, an array of 1.5 mm wide, 28 μm thick PVDF films were positioned 0.5 mm apart. A picture of a sensor with seven sensing elements is shown in Figure 6.10. To apply uniform distributed load, a flat plate was attached to the probe, as shown in Figure 6.11. The experimental setup has been explained in previous chapters and is shown in Figure 3.23. The outputs of the sensors were fed into the connector box through an interface circuit.

The data was then transferred to the computer using a DAQ (NI PCI-6225, National Instruments). To reduce the noise effect in each experiment, the average of the peak values

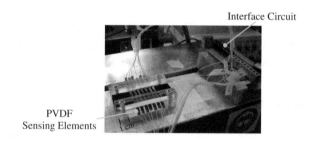

Figure 6.10 A sensor with seven sensing elements is connected to the DAQ (See Plate 13)

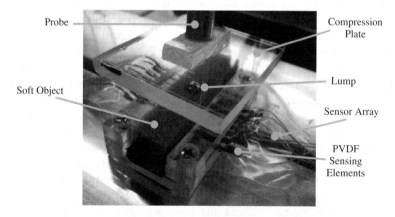

Figure 6.11 Soft object with embedded lumps under test. The top plate was used to replicate the upper jaw of the grasper in order to apply compressive loads

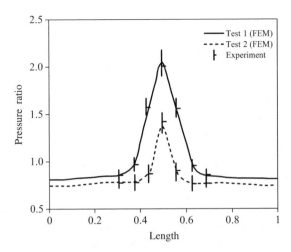

Figure 6.12 The experimental data versus finite element results (error bar indicates the range of readings). The simulation conditions for Test 1: U_y = 2.5 mm, R = 3 mm, and Depth = 5 mm; and for Test 2 : U_y = 1 mm, R = 2 mm, and Depth = 4 mm

of at least 20 cycles were recorded and arithmetically averaged. In the first test (Test 1), the parameters were as follows: applied displacement U_y = 2.5 mm; diameter and depth of the lump were 6 and 5 mm, respectively. Whereas in the second test (Test 2), the applied displacement was U_y = 1 mm with a lump diameter and depth of 4 mm. The experimental results are compared with finite element results and shown in Figure 6.12.

6.6 Discussion and Conclusions

The results obtained from the finite element analysis of a mass embedded in soft tissue are presented in Section 6.4 and show that the pressure distribution at the contact surface is influenced by several parameters, such as size, depth, and stiffness of the lump, as well as the applied force. The nonlinearity of the response with respect to these parameters is also shown. This nonlinearity can be attributed to both the material nonlinearity of soft tissue and geometrical nonlinearity, which happens for large strains. The response to variation in applied load, as depicted in Figure 6.7, shows that although an increase in grasping load causes an increase in the maximum pressure, the shape of curve remains unchanged. In addition, the amount of area that each curve covers remains almost constant. This could be a distinctive sign that differentiates the effect of variations in load from the variations in the size and depth of the lump. It is shown that although the ratio of the Young's modulus of the lump to that of the tissue is an effective factor, the stress profile remains almost unaltered for ratios higher than 10. On the other hand, the pressure ratio is highly sensitive to the variation in stiffness of the lump for $E_L < 10\, E_T$, which potentially is helpful in detecting lumps in their early stages.

The thickness of tissue clearly affects the stress profile; however, for a specific arrangement of lump and tissue, the increase in tissue thickness is equal to the decrease in lump size. Therefore, it is possible to combine both parameters into one variable, in order to reduce the number of involved parameters, by introducing a dimensionless ratio D/t, the lump diameter to the tissue thickness.

Soft tissue shows very complex behavior, including hysteresis and viscoelasticity, which were not considered in this study. Nevertheless, special attention was paid to modeling accurately the nonlinearity of the tissue. In order to accomplish this, an elastomer was mechanically tested and the experimental data used for simulations. Initially several strain energy functions, such as, Ogden, Neo–Hookean, and Mooney–Rivlin were considered and finally it was found that the three-term Mooney–Rivlin function concurred favorably with the experimental data.

The assumptions made in the geometry and boundary conditions were that they did adequately replicate the state of the tissue grasped by a typical MIS or surgical robot grasper tool and that, in either case, the existence of a lump within the grasped tissue affected the pressure distribution on the contact surface. Therefore, close inspection of the pressure distribution reveals valuable information about the lump, which is useful, particularly in developing inverse models of the problem, as well as in designing smart endoscopic graspers. The findings of this study can be used to calculate the required sensitivity and resolution of those sensing elements located at the contact surface, as well as the spatial resolution of any proposed array or matrix of the sensors for this purpose. In addition, the tactile image can be used for the visualization and enhancement of existing 2D video images.

References

1. Miyaji, K., Furuse, A., Nakajima, J. *et al.* (1997) The stiffness of lymph nodes containing lung carcinoma metastases. *Cancer*, **80**, 1920–1925.
2. Barman, I. and Guha, S.K. (2006) Analysis of a new combined stretch and pressure sensor for internal nodule palpation. *Sensors and Actuators A: Physical*, **125**, 210–216.
3. Hosseini, M., Najarian, S., Motaghinasab, S., and Dargahi, J. (2006) Detection of tumours using a computational tactile sensing approach. *The International Journal of Medical Robotics and Computer Assisted Surgery*, **2**, 333–340.
4. Wellman, P.S., Dalton, E.P., Krag, D. *et al.* (2001) Tactile imaging of breast masses: first clinical report. *Arch Surg*, **136**, 204–208.
5. Wellman, P.S. and Howe, R.D. (1999) Extracting features from tactile maps. Proceedings of Medical Image Computing and Computer-Assisted Intervention, 1999, pp. 1133–1142.
6. Wellman, P.S., Howe, R.D., Dewagan, N. *et al.* (1999) Tactile imaging: a method for documenting breast masses. Proceedings of the First Joint, BMES/EMBS Conference, 1999 Engineering in Medicine and Biology 21st Annual Conference and the 1999 Annual Fall Meeting of the Biomedical Engineering Society, p. 1131.
7. Kattavenos, N., Lawrenson, B., Frank, T.G. *et al.* (2004) Force-sensitive tactile sensor for minimal access surgery. *Minimally Invasive Therapy & Allied Technologies*, **13**, 42–46.
8. Gerhard A. Holzapfel. Nonlinear Solid Mechanics: A Continuum Approach for Engineering. John Wiley & Sons Ltd., Baffins Lane, Chichester, West Sussex PO19 1UD, England, 2000.
9. ANSYS Inc. (2004) Theory Reference, ANSYS Release 9.0.002114.
10. Williams, J.G. (1973) *Stress Analysis of Polymers*, Longman Group Limited, London.
11. Karol, M. (2000) Constitutive modelling of abdominal organs. *Journal of Biomechanics*, **33**, 367–373.
12. ANSYS Inc. Documentation for ANSYS, Modeling Material Nonlinearities/Mixed u-p Formulation, Release 10.

13. Sussman, T. and Bathe, K.-J. (1987) A finite element formulation for nonlinear incompressible elastic and inelastic analysis. *Computers and Structures*, **26**, 357–409.
14. Bonet, J. and Wood, R.D. (1997) *Nonlinear Continuum Mechanics for Finite Element Analysis*, Cambridge University Press.
15. Crisfield, M.A. (1997) *Non-linear Finite Element Analysis of Solids and Structures*, Advanced Topics, Vol. **2**, John Wiley & Sons, Inc.
16. Bathe, K.J. (1996) *Finite Element Procedures*, Printice-Hall.
17. McMeeking, R.M. and Rice, J.R. (1975) Finite-element formulations for problems of large elastic–plastic deformation. *International Journal of Solids and Structures*, **11**, 601–616.
18. Holzapfel, G.A. (2000) *Nonlinear Solid Mechanics: A Continuum Approach for Engineering*, John Wiley & Sons Ltd.
19. Rivlin, R.S. and Eirich, F.R. (1956) Large elastic deformations. *Rheology: Theory and applications*, **24**, 351–385.
20. Winnie, Y., Yongbao, L., Zheng, Y.P. *et al*. (2006) Softness measurements for open-cell foam materials and human soft tissue. *Measurement Science and Technology*, **17**, 1785–1791.

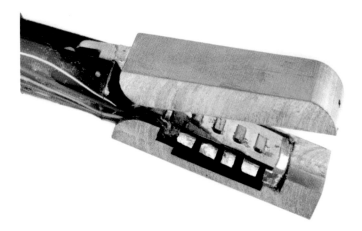

Plate 1 Photograph of the prototyped endoscopic grasper integrated with the tactile sensor

Plate 2 A proposed design of the smart grasper in which the incorporation of microfabricated sensor is presented

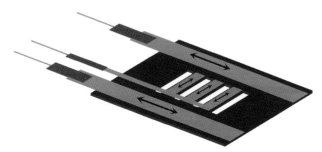

Plate 3 Backside view of the sensor's top part. The positions of the PVDF films and the electrode configurations are illustrated in this view. Arrows on the PVDF film strips indicate the orientation of the drawn direction of the PVDF film

Tactile Sensing and Displays: Haptic Feedback for Minimally Invasive Surgery and Robotics, First Edition.
Saeed Sokhanvar, Javad Dargahi, Siamak Najarian, and Siamak Arbatani.
© 2013 John Wiley & Sons, Ltd. Published 2013 by John Wiley & Sons, Ltd.

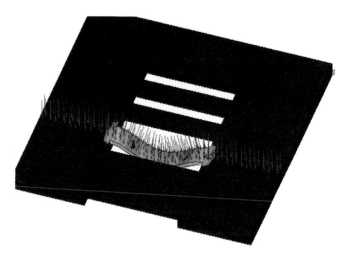

Plate 4 The deflected structure when a distributed load of 111 kPa was applied to the first beam and its two supports

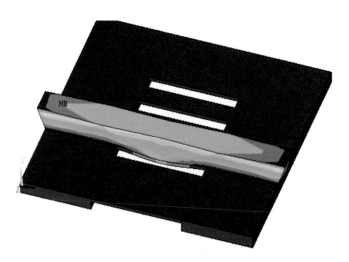

Plate 5 The finite element model of the sensor when a soft material is placed on the first tooth and a uniform compressive load is applied

■■■ Silicon Oxide
■■■ Photo Resist
▨▨▨ Protection Layer

Step	Description	Specification	Schematic
1-1	Pre-Cleaning	10 min in acetone (in ultrasonic bath) 10 min in iso-propanol (in ultrasonic bath)	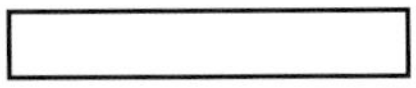
1-2	Cleaning	10 min @ Piranah	
2	Removing Oxide Layer	20 Sec @ HF, 1%	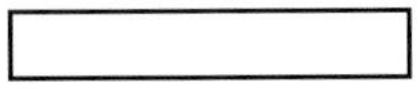
3	Wet Oxidation	20 min for 1000 Å	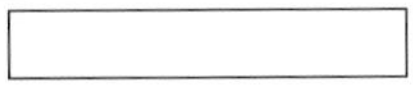
4-1	Spin Coating of Photoresist	30 sec @ 4000 rpm	
4-2	Soft Baking	1 min @ 115 °C	
4-3	Exposure	5 sec, a 270 watts mercury lamp	
4-4	Developing	1 min	
4-5	Hard Baking	30 min @ 105 °C	
5	Backside Protection	Using special scoth tape	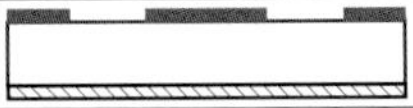
6	Oxide Layer Etching	2 min in BOE, 7:1	
7	Removing Backside Protection		
8	Photoresist Strip off	10 min in acetone (in ultrasonic bath) 10 min in iso-propanol (in ultrasonic bath)	
9	TMAH etching	1.5 hr for bottom layer (TMAH 25% @90 °C) 4.5 hr for top laye (TMAH 25% @90 °C)	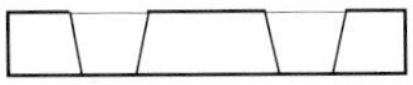
10	Removing Oxide Layer	3 min in buffered HF @ room temperature	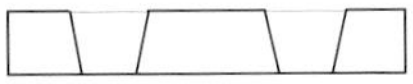
11	Wet Oxidation	20 min for 1000 Å	

Plate 6 The sequence of micromachining process summarized

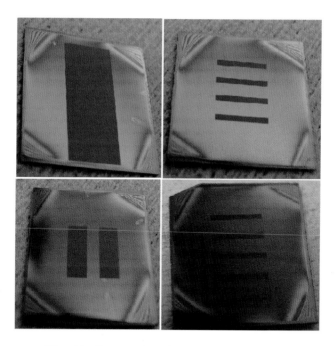

Plate 7 The samples after development process

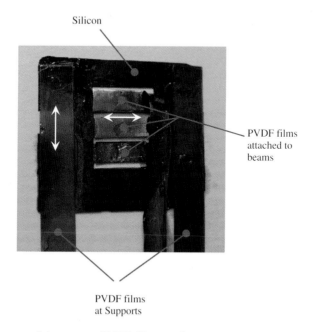

Plate 8 The top part of the sensor. PVDF films at the supports as well as films attached to the beams are shown. The direction of the PVDF films are also illustrated

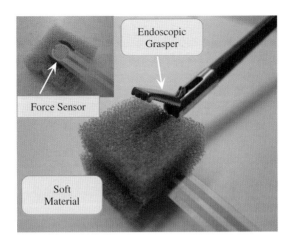

Plate 9   A typical grasper jaw, elastomeric material, and FSR sensor

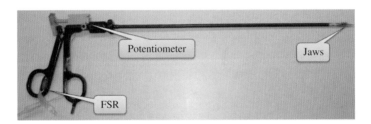

Plate 10   A picture taken from endoscopic grasper equipped with two FSR sensors and a linear potentiometer

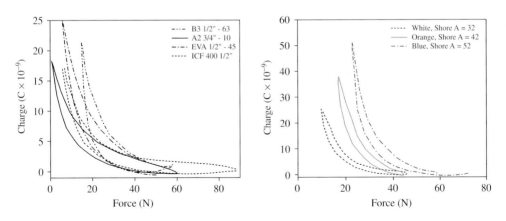

Plate 11   The force–angle curve for (a) ICF400, EVA, B3, and A2 and (b) silicone rubbers (white, Shore A=32; orange, Shore A=42; blue, Shore A=52)

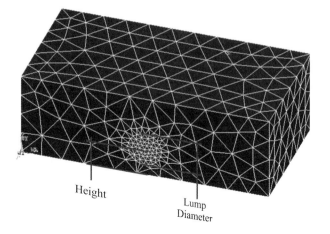

Plate 12 A half-model of soft tissue which contains a lump

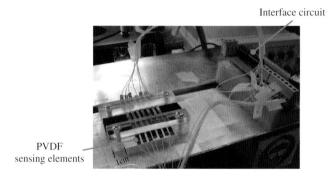

Plate 13 The sensor with seven sensing elements is connected to the DAQ

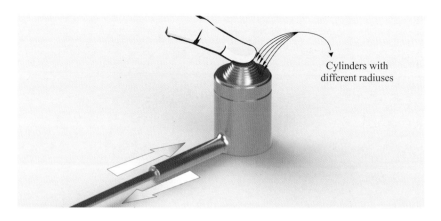

Plate 14 Softness display prototype

Plate 15 Electro-rheological fluid at reference (left) and activated states (right) [20] (© Institute of Physics and IOP Publishing)

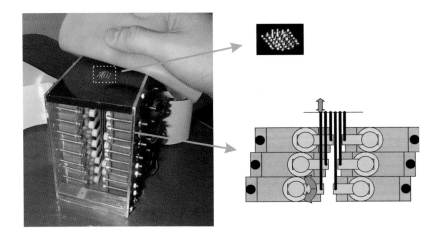

Plate 16 A tactile shape display using servomotros

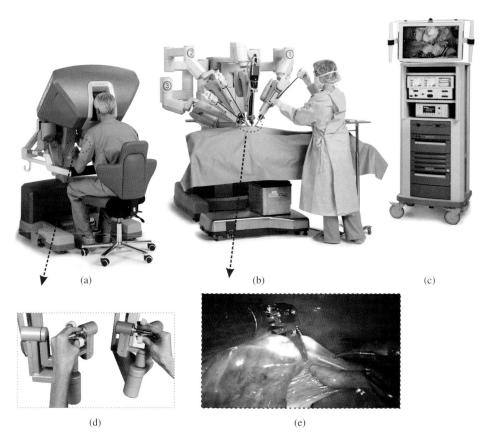

(a)　　　　　　　　　(b)　　　　　　　　　(c)

(d)　　　　　　　　　(e)

Plate 17 The daVinci telesurgical robot extends a surgeon's capabilities by providing the immediacy and dexterity of open surgery in a minimally invasive surgical environment. (Photos: Intuitive Surgical, Sunnyvale)

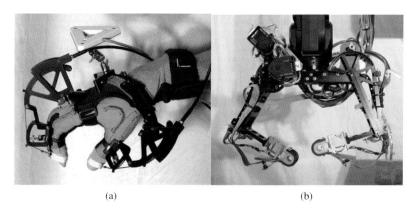

(a)　　　　　　　　　　　(b)

Plate 18 Dextrous Telemanipulation system, of Stanford University. a) The master system consisting of an instrumented glove for finger motion measurement and an exoskeleton for fingertip force feedback. b) The slave robotic hand with two fingers and fingertip force sensors for relaying environmental interactions

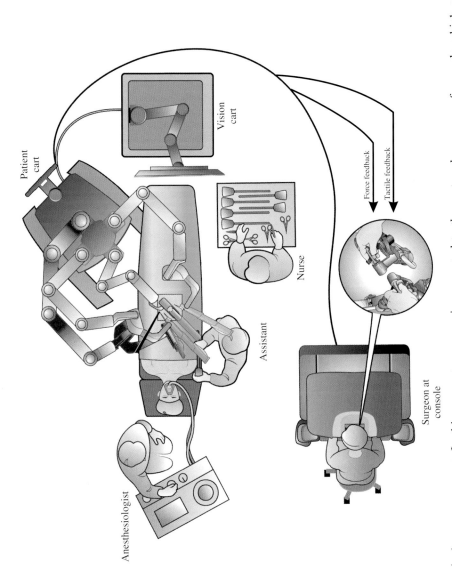

Plate 19 Telesurgical system concept. In this concept, surgeon is connected to the system by means of console which provides the surgeon with vision, and master controls which translates the surgeon's hand, wrist and finger movements into precise, real-time movements of surgical instruments. The robotic arms in the patient cart are equipped with surgical instruments and lets every surgical maneuver be under the direct control of the surgeon

Patient cart

Vision cart

Nurse

Assistant

Anesthesiologist

Force feedback

Tactile feedback

Surgeon at console

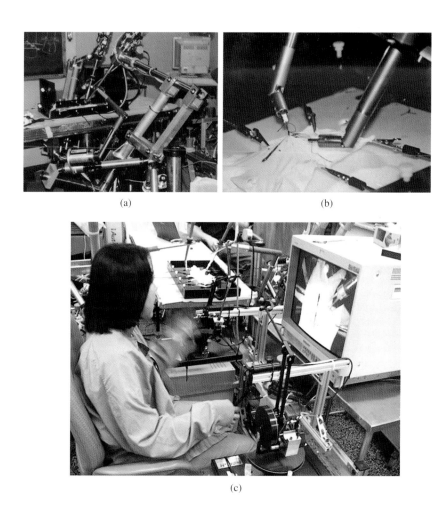

(a)

(b)

(c)

Plate 20 (a) and (b) Slave manipulator of the University of California, Berkeley and University of California, San Francisco laparoscopic telesurgical workstation, tying a knot in the training box. (c) Master workstation of the Robotic Telesurgical Workstation (RTW)

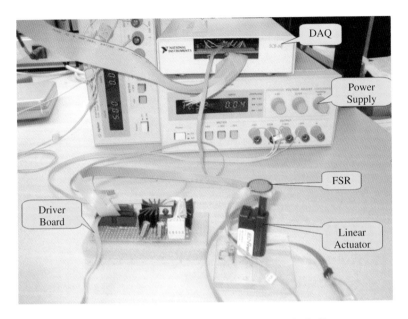

Plate 21 Photograph of the tactile display including

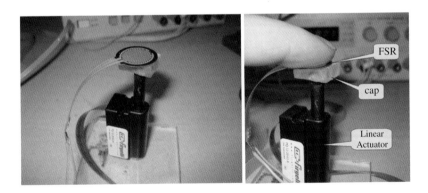

Plate 22 The linear actuator, FSR and plexiglas cap

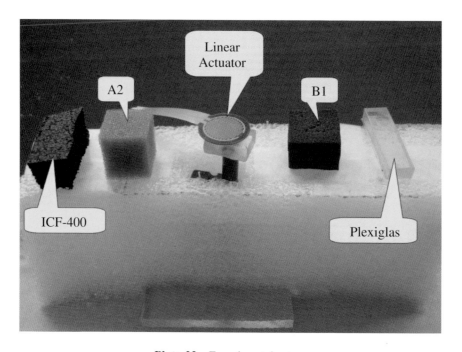

Plate 23 Experimental setup

7

Tactile Display Technology

A tactile device is a man–machine interface that can reproduce accurately tactile parameters such as texture, roughness, temperature, and shape, and is used in virtual environments (VEs) (applications based on virtual reality technology) and teleoperation applications [1, 2].

This chapter provides an overview as to what has been achieved so far in this field and to summarize the different specifications that may contribute to the development and evolution of future tactile interface and display systems.

The following list classifies the domain and possible applications for various tactile interfaces:

- teleoperation and telepresence;
- laboratory prototypes to study different tactile parameters;
- sensory substitution;
- 3D surface generation;
- Braille systems;
- games.

From a technological point-of-view, tactile stimulation can be accomplished in a number of different ways. Technologies used for VE systems were inspired by dot matrix printers technologies and Braille systems for the blind. Solutions have been proposed based on mechanical needles actuated by electromagnetic technologies (solenoids, voice coils), piezoelectric crystals, shape-memory alloys (SMAs), pneumatic systems, and heat pump systems based on Peltier modules and dielectric elastomers.

Other technologies that use electrorheological fluid (ERF) are still under investigation. When subjected to an electrical field, this fluid changes its consistency from thick to thin within milliseconds, making it useful in many different hydraulic actuator applications.

These, and other technologies shown in Figure 7.1, can be used in the field of tactile sensing to provide information on various sensations, such as vibration, shear, heat, pressure, friction, scratch, and indentation [3].

Tactile Sensing and Displays: Haptic Feedback for Minimally Invasive Surgery and Robotics, First Edition.
Saeed Sokhanvar, Javad Dargahi, Siamak Najarian, and Siamak Arbatani.
© 2013 John Wiley & Sons, Ltd. Published 2013 by John Wiley & Sons, Ltd.

Technologies   Interaction Modes

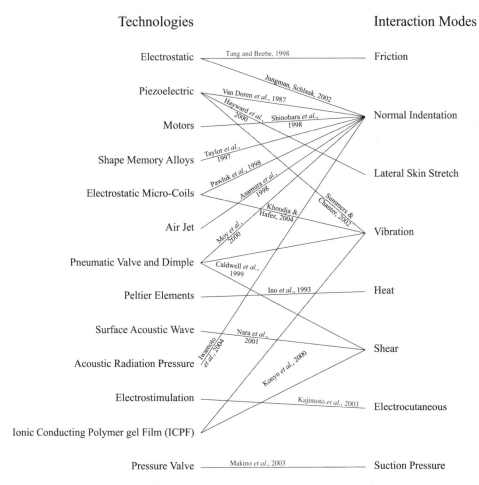

Figure 7.1 Examples of combinations of modes of interaction and actuator technologies for the design of tactile displays [3]

7.1 The Coupled Nature of the Kinesthetic and Tactile Feedback

The nature of haptic systems, together with their complex integrated modality, has been investigated by many researchers. Khatchatourov *et al.* [4] believe that the tactile, kinesthetic, and thermal aspects of the human haptic system are closely connected and that, furthermore, a tactile/force-feedback relationship is linked by a reciprocal theory which basically states that any behavior influences, and is influenced by, personal factors and the social environment, whereby spatial aspects of haptic modality is, itself, a composite of both the tactile and proprioceptive modalities. They showed the necessity of integrating force and tactile feedbacks in order to address the many limitations of devices which only have either tactile or force feedback capabilities. They targeted conceptual, technological, and experimental approaches in order to address the complexity of integrating force and tactile feedback systems by modeling highly rigid or very soft interactions,

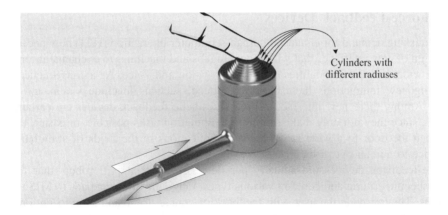

Cylinders with
different radiuses

Figure 7.2   Softness display prototype [6] (See Plate 14)

using a computer model and a tactile force-feedback device (ERGOS [5]). Their experiments illustrated that the addition of tactile feedback (virtual finger deformability in this investigation) to force feedback, increases both accuracy and the number of tasks accomplished (in this investigation, following the task, they modeled a virtual bridge contour for evaluation of feedback accuracy).

In work conducted by Scilingo *et al*. [6, 7], a combination of a kinesthetic and tactile display was proposed and developed. This system is capable of independent and accurate rendering of both kinesthetic and cutaneous cues in order to provide perception and precise discrimination of softness. In their early work [8], they illustrated that most of the cutaneous information can be derived from contact force and contact area (spread rate of contact area). In order to render softness, they proposed a novel softness display [8] consisting of a pneumatic device that contained a set of cylinders with different radii, as shown in Figure 7.2, and combined this in series with a commercial force-feedback interface, the Delta Haptic Device (DHD). In this system, they assigned two independent control inputs (chamber pressure and DHD force) to the cutaneous and kinesthetic parts of their simplified system.

Mechanical interaction during contact causes several complex stimulations of receptors in the skin and in the proprioceptive system [9]. To illustrate the complexity of contact, and the need for integrating kinesthetic and cutaneous cues, Scilingo *et al*. [6] proposed an experiment to show how two objects having an equal kinesthetic cue can present different cutaneous cues, and vice versa. To assess the impact of integration and performance of their proposed integrated display, they conducted psychophysical experiments, and compared the subjects' indirect perceptions of softness with those obtained by direct touch on physical objects. It can be observed from their results that the subjects interacting with the integrated haptic display were able to discriminate the softness better than by using either purely kinesthetic or cutaneous displays. By referring to Srinivasan and LaMotte's work [10], it can be concluded that for compliant objects with rigid surfaces, a combination of both tactile and kinesthetic information (active touch) is essential, otherwise, if only one or the other were used, maximum discrimination would not be possible, as indicated by Lederman and Klatzky [11].

7.2 Force-Feedback Devices

The increasing demand for enhanced human–computer interaction (HCI) now necessitates the design of new interfaces that will allow humans and machines to exchange information using a wider range of modalities than merely vision and voice. As a consequence, ideas involving new interactions have been introduced, such as machine vision and virtual reality. Among these new interfaces, force kinesthetic feedback devices are certain to be favored, since they not only make difficult manipulation tasks possible, or easier, but they also open the door to a wide range of new applications in the fields of stimulation and assistance to human operators.

Force-feedback devices are actually parallel [12] or serial [13] robots that function within specific criteria and come in various types, ranging from gamepads to MIS training facilities. They provide the user with feedback of contact force and proprioception of the manipulator hand, or virtual object, during an interaction. In these kinds of robots, the mechanical system should have high stiffness, low inertia, and friction with no backlash. Also, force actuators should be capable of back-drivability and be provided with sufficient force/torque precision. The most precise position sensors should be employed in a high-frequency local control loop (greater than 1 kHz) to make the vibrations sensible.

Up to now, commercially successful force-feedback products have been mainly centered on the entertainment industry, since, unfortunately, their use in the medical sector has been severely curtailed due to the fact they are unable to provide cutaneous cues. Notwithstanding that, commercial force feedback devices have achieved outstanding results in displaying hardness within their natural impedance range. Softness displays have proved that they are able to evoke a reliable softness sensation, enabling better discrimination than that of similar, but purely kinesthetic, displays. On the other hand, pure softness displays have limited workspace and softness range, and it is impossible to decouple the cutaneous and kinesthetic information purely by using tactile displays.

7.3 A Review of Recent and Advanced Tactile Displays

Hitherto, the technologies pertaining to conventional kinesthetic and tactile displays, together with the conceptual aspects of their integration, were discussed in some detail. This section introduces some of the more recent and advanced tactile displays.

7.3.1 Electrostatic Tactile Displays for Roughness

Pelrine *et al.* reported on an electrostatic actuator composed of a polymeric elastic dielectric that is sandwiched between compliant electrodes [14]. By applying a voltage difference between the electrodes, the dielectric contracts in thickness and expands its area due to the attractive charges on the electrodes [15] (see Figure 7.3). By reducing the voltage, the dielectric returns to its initial shape and can produce forces due to the stored elastic energy.

In order achieve this effect, given the applicable voltages, the dielectric can be no more than a few microns thick and the electrodes must be very compliant in order that they do not constrain the deformation. Pelrine *et al.* showed more than 30% relative strain in thickness at electrostatic pressures of over 1 MPa [14] with a silicone elastomer dielectric

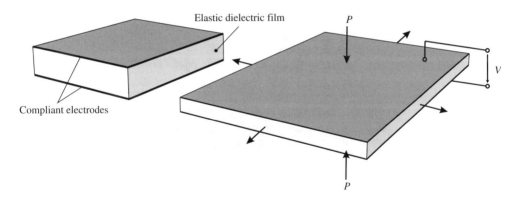

Figure 7.3   Deformation of an elastic dielectric film under electrostatic pressure [15]

between carbon electrodes. Jungmann and Schlaak [15] used a stack of many layers of dielectric films and electrodes to obtain the large absolute displacements depicted in Figure 7.4.

If a voltage (V) is applied between the neighboring electrode layers, the actuator stack will contract and the stimulator tip will disappear below the surface of the device. To ensure the area expansion, there has to be a gap around the actuator stack. By reducing the voltage, the electrodes are discharged beyond the voltage source, causing a relaxation of the actuator stack. As a result, the stimulator tip is pressed against the skin on top of the device because of the stored elastic energy.

A stimulator with a relative strain of 30% at an absolute value of 4 mm would require more than 1000 dielectric layers, 10 mm in thickness. This shows the necessity of automating the actuator processing [15]. In Figure 7.5, a possible stimulator arrangement for a tactile display with elastomeric actuators is shown.

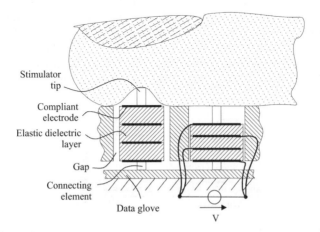

Figure 7.4   Structure and function of an electrostatic tactile stimulator with elastic dielectric [15]

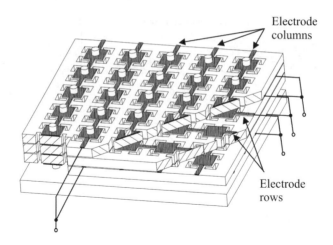

Figure 7.5   Planar stimulator arrangement for a tactile display [15]

Yamamoto worked on a tactile display in which electrostatic force and friction control were developed for presenting surface roughness sensations [16–18]. The device consists of stator electrodes and a thin film slider upon which an aluminum conductive layer is deposited. The user puts his index finger on the slider and moves it horizontally to obtain a certain tactile sensation. By applying various voltage patterns to stator electrodes, various friction distributions can be generated on the slider which, in turn, is transferred to the fingertip, so as to generate surface roughness sensation.

7.3.2   Rheological Tactile Displays for Softness

ERFs experience dramatic changes in rheological properties, such as viscosity, in the presence of an electric field. This was first explained by Willis Winslow in 1947 using oil dispersions of fine powders [19]. The fluids are made from suspensions of an insulating base and particles of the order of 0.1 to 100 µm (in size). The volume fraction of the particles is between 20 and 60%. The electrorheological effect, sometimes called the Winslow effect, is thought to arise from the difference in the dielectric constants of the fluid and particles. In the presence of an electric field, due to an induced dipole moment, the particles will form chains along the field lines, as shown in Figure 7.6.

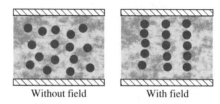

Without field   With field

Figure 7.6   Particle suspension forms chains when an electric field is applied [20] (© Institute of Physics and IOP Publishing)

Figure 7.7   Electrorheological fluid in reference (left) and activated states (right) [20] (© Institute of Physics and IOP Publishing) (See Plate 15)

These chains alter the ERF viscosity, yield stress, and other properties, allowing it to change consistency from that of a liquid to something that is viscoelastic, such as a gel, with response times to changes in electric fields on the order of milliseconds. Figure 7.7 shows the fluid state of an ERF without an applied electric field and the solid-like state (i.e., when an electric field is applied). Good reviews of the ERF phenomenon and the theoretical basis for ERF behavior can be found in [21–24].

Using rheological fluids does have some disadvantages, however. For example, it is considered a hazardous substance and the user must avoid direct contact with the fluid since it contains oxalic acid. Another problem is that because ERFs are suspensions, they tend to settle out in time. Finally, it is very difficult to get good spatial resolution in any actuating device.

7.3.3   Electromagnetic Tactile Displays (Shape Display)

This type of tactile display uses linear movements of an electromotor to simulate the shape. A good example of this category is the work achieved by Wagner *et al*. [25] in which they used an array of 36 (6 × 6) servomotors as a shape display. Figure 7.8 shows the pin array and the servomotors, which are arranged in such a way that, as shown in this figure, their rotational movements are converted to linear movements of the pins.

Ottermo *et al*. worked on fabricating a shape display using micromotors on the handle of an endoscopic grasper. The sensor array is 24 × 8 mm and consists of 30 piezoelectric sensors, while the tactile display comprises 30 micro motors adding up to a total size of 32 × 18 × 45 mm [26, 27].

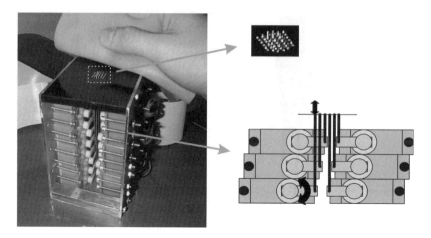

Figure 7.8   A tactile shape display using servomotors [25] (See Plate 16)

7.3.4   Shape Memory Alloy (SMA) Tactile Display (Shape)

SMA wire shortens when it undergoes a phase transition from the Martensitic to the Austenitic phase. This reversible phase transition can be produced by passing current through the wire and heating it to above its transition temperature. It is well known that SMA exhibits hysteresis when cycled through this transition [28]. It is also a relatively slow process because it can take a long time for the wire to cool down and return to its original length. Some researchers have suggested that it is possible to increase the bandwidth and account for the hysteresis by developing a model of the process and incorporating this into the controller [29, 30].

While this undoubtedly will work to some degree, Wellman *et al.* have chosen to increase the bandwidth through careful thermal design, and have minimized the hysteresis through position feedback control of the display [31]. Figure 7.9 shows how a single pin is actuated by the SMA. A 3D shape display, using an array of bars actuated by SMA, is presented in [32–34] and shown in Figure 7.10.

7.3.5   Piezoelectric Tactile Display (Lateral Skin Stretch)

Piezoceramic actuators constitute a practical choice to build an actuator array because they are based on the phenomenon of deformation of a quartz crystal caused by an electrical field. They can be operated over a large bandwidth and are relatively easy to form into a desired miniature structure. Moreover, they are widely available from a variety of sources and are relatively inexpensive. Unfortunately, piezoceramic actuators still require high operating voltages although, with the advent of technology, they are steadily improving and their usefulness is becoming more apparent with time. Hayward *et al.* worked on a tactile display using an array of piezoceramics [35–38]. The objective was to create a deformable structure upon whose exposed side there was to be an array of contactors, each of which would contact a patch of skin in a tangential motion and produce programmable strain fields. Among the different piezoelectric elements available, bimorphs can achieve

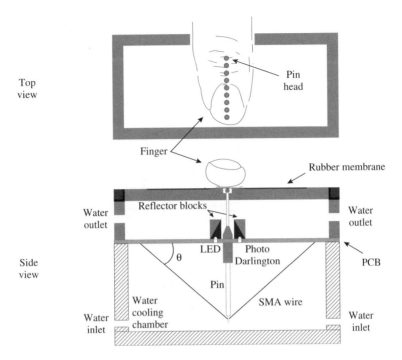

Figure 7.9 Tactile display with SMA in a V-shaped configuration [31 (© Springer)]

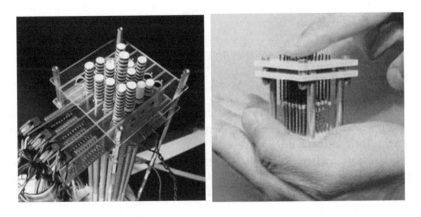

Figure 7.10 3D shape displays [32, 33]

substantial displacement by bending a cantilever (Figure 7.11a). This bending motion can directly be used to stretch the skin without the need for extra motion amplification mechanisms. Figure 7.11b shows the manner in which a collection of bending elements can be arranged to create a two-dimensional array of contactors. One advantage of this design is the creation of a strong, yet modular, structural configuration made of the same part replicated 10 times. Although all contactors move along the same direction due to the

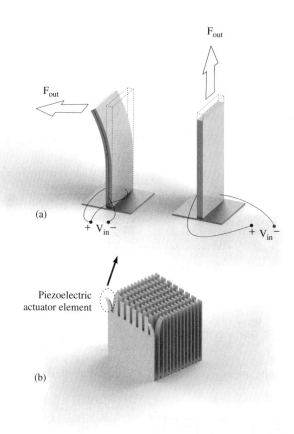

Figure 7.11   (a) Bending of a bimorph and (b) bending elements arranged into an array [36–38]

fact that all the actuators bend around the same axis, this was not found to be a limitation since, at such a small scale, the resulting sensations seem to be indifferent to the details of skin stretch/shear orientation.

In another research work, Kwon *et al*. described the development of a piezoelectric-based planar distributed tactile display capable of displaying textures [35, 39]. The tactile display is composed of a 6 × 5 pin array actuated by 30 piezoelectric bimorphs.

7.3.6   Air Jet Tactile Displays (Surface Indentation)

A new tactile stimulation method is proposed in [40–42] by controlling the suction pressure. This method is based on our discovery of a tactile illusion where we feel as if something like a stick is pushing up into the skin surface when we pull skin through a hole by lowering the air pressure.

This illustration shows how our tactile mechanoreceptors detect strain energy, but cannot differentiate between positive or negative stress. Therefore we are unable to discriminate suction from compression (see Figure 7.12a,b).

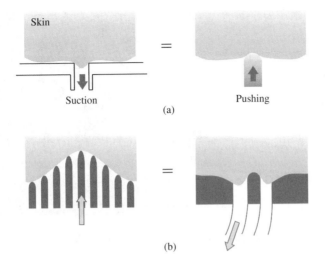

Figure 7.12   We feel a pushing sensation from suction pressure [40]

7.3.7   Thermal Tactile Displays

Thermomechanical actuators provide good performance in terms of displacement, force, and work per cycle. Thermopneumatic micropumps and microvalves are based on sealed cavities that have a flexible side [43]. The cavities are filled with a low boiling point liquid (for instance methyl chloride) and a resistive heater is built inside.

When the heater increases the temperature in the cavity, the pressure grows because of gas resulting from the liquid–gas phase transition and the flexible side of the cavity is displaced. Figure 7.13 shows the actuator proposed in [44]. It consists of a small cylinder made of copper with one end sealed with tin and the other with a flexible diaphragm. A signal diode has been chosen as the heater due to its small size, although it can be

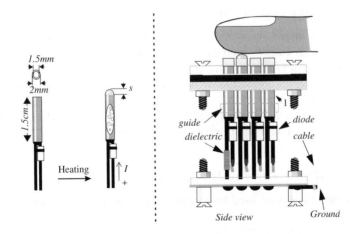

Figure 7.13   Thermomechanical actuator [44]

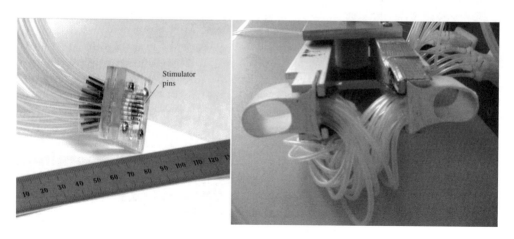

Figure 7.14 Pneumatic tactile display

replaced by a resistor or another semiconductor device. The main application of the thermal tactile display is in Braille cells [45].

7.3.8 Pneumatic Tactile Displays (Shape)

To provide local shape information, an array of force generators can create a pressure distribution on a fingertip, thereby synthesizing an approximate true contact. Researchers at the University of Berkley [46, 47] have developed a prototype of a 5 × 5 tactile interface actuated by pneumatic technology with 3 bits of resolution. The pneumatic prototype interface delivers up to 0.3 N per actuator. This tactile display, which is similar to a servomotor tactile display, has certain advantages, such as ease of fabrication and no pin friction. Figure 7.14 shows the pin array and the molding which covers the pins.

7.3.9 Electrocutaneous Tactile Displays

An electrocutaneous display is a tactile device that directly activates nerve fibers within the skin with electrical current from surface electrodes (Figure 7.15), thereby generating sensations of pressure or vibration without the use of any mechanical actuator [16, 48–50]. It is known that the nerves connected to the tactile receptors can be stimulated selectively by electrostimulation to some extent. Meissner's corpuscles, which detect the dynamic deformation of a finger, can be stimulated by anode current, while Merkel's disks, which detect static deformation, can be stimulated by cathode current [51, 52].

7.3.10 Other Tactile Display Technologies

Other technologies have been used to develop tactile displays, including:

- Producing tactile sensation with acoustic radiation pressure [53].
- Tactile glove that exhibits pressure and temperature feedback, called Teletact [54].
- Effect of linear combination of magnets on the skin [55].

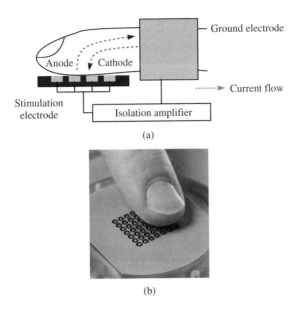

(a)

(b)

Figure 7.15   (a) Electric current through the finger stimulates nerves and (b) an electrocutaneous tactile display

- Array of vibrating pixels using micro-clutch micro-electro-mechanical systems (MEMS) technology [56].
- Using a surface acoustic wave (SAW) [57].
- Fusion of visual and tactile feedback [58].
- Composite polymeric known as IPMC or ICPF [59]
- Digital clay using MEMS technology [60].

References

1. Fearing, R.S. (1990) Tactile sensing mechanisms. *The International Journal of Robotics Research*, **9**, 3–23.
2. Cohn, M.B., Lam, M., and Fearing, R.S. (1992) Tactile feedback for teleoperation. Proceedings of SPIE 1992 Telemanipulator Technology, vol. 1833, pp. 240–254.
3. Pasquero, J. (2006) Survey on Communication through Touch. Technical Report TR-CIM, vol. 6. Center for Intelligent Machines-McGill University.
4. Khatchatourov, A., Castet, J., Florens, J.L. *et al.* (2009) Integrating tactile and force feedback for highly dynamic tasks: technological, experimental and epistemological aspects. *Interacting with Computers*, **21**, 26–37.
5. Florens, J.L., Luciani, A., Cadoz, C., and Castagné, C. (2004) ERGOS: multi-degrees of freedom and versatile force-feedback panoply. Proceedings of EuroHaptics, 2004, pp. 356–360.
6. Scilingo, E.P., Bianchi, M., Grioli, G., and Bicchi, A. (2010) Rendering softness: integration of kinesthetic and cutaneous information in a haptic device. *IEEE Transactions on Haptics*, **3**, 109–118.
7. Scilingo, E.P., Sgambelluri, N., Giovanni, T., and Bicchi, A. (2007) Integrating two haptic devices for performance enhancement. EuroHaptics Conference, 2007 and Symposium on Haptic Interfaces for Virtual Environment and Teleoperator Systems, Tsukaba, 2007, pp. 139–144.
8. Ambrosi, G., Bicchi, A., Rossi, D.D., and Scilingo, E.P. (1999) The role of contact area spread rate in haptic discrimination of softliess. IEEE International Conference on Robotics and Automation, Detroit, MI, 1999.

9. Johnson, K.O. (2001) The roles and functions of cutaneous mechanoreceptors. *Current Opinion in Neurobiology*, **11**, 455–461.
10. Srinivasan, M.A. and LaMotte, R.H. (1996) Tactual discrimination of softness: abilities and mechanisms, *Somesthesis and the Neurobiology of the Somatosensory Cortex*, Birkhauser Verlag AG, pp. 123–135.
11. Lederman, S.J. and Klatzky, R.A. (2004) Haptic identification of common objects: effects of constraining the manual exploration process. *Perception & Psychophysics*, **66**, 618–628.
12. Clavel, R. and Burckhardt, C.W. (1991) *Conception d'un Robot Parallèle Rapide à 4 degrés de Liberté*, LSRO, EPFL.
13. Beckman, J.A. (2007) The PHANTOM Omni as an Under-Actuated Robot. M.S. Aerospace Engineering, Iowa State University, United States-Iowa.
14. Pelrine, R., Kornbluh, R., Joseph, J. *et al*. (2000) High-field deformation of elastomeric dielectrics for actuators. *Materials Science and Engineering: C*, **11**, 89–100.
15. Jungmann, M. and Schlaak, H.F. (2002) Miniaturised electrostatic tactile display with high structural compliance. Proceedings of the Eurohaptics, Edinburgh, UK, 2002.
16. Yamamoto, A., Nagasawa, S., Yamamoto, H., and Higuchi, T. (2006) Electrostatic tactile display with thin film slider and its application to tactile telepresentation systems. *IEEE Transactions on Visualization and Computer Graphics*, **12**, 168–177.
17. Yamamoto, A., Ishii, T., Takasaki, M. *et al*. (2001) Tactile interface using ultra-thin electrostatic actuator. Proceedings of the 32nd ISR (International Symposium on Robotics), 2001, pp. 19–21.
18. Yamamoto, A., Ishii, T., and Higuchi, T. (2003) Electrostatic tactile display for presenting surface roughness sensation. IEEE International Conference on Industrial Technology, 2003, vol. 2, pp. 680–684.
19. Winslow, W.M. (1949) Induced fibrillation of suspensions. *Journal of Applied Physics*, **20**, 1137–1140.
20. Weinberg, B., Nikitczuk, J., Fisch, A., and Mavroidis, C. (2005) Development of electro-rheological fluidic resistive actuators for haptic vehicular instrument controls. *Smart Materials and Structures*, **14**, 1107.
21. Weiss, K., Carlson, D., and Coulter, J. (1993) Material aspects of electrorheological systems. *Journal of Intelligent Material Systems and Structures*, **4**, 13–34.
22. Gast, A.P. and Zukoski, C.F. (1989) Electrorheological fluids as colloidal suspensions. *Advances in Colloid and Interface Science*, **30**, 153–202.
23. Conrad, H. (1998) Properties and design of electrorheological suspensions. *MRS Bulletin*, **23**, 35–42.
24. Block, H. and Kelly, J.P. (1988) Electro-rheology. *Journal of Physics D: Applied Physics*, **21**, 1661.
25. Wagner, C.R., Lederman, S.J., and Howe, R.D. (2002) A tactile shape display using RC servomotors. Proceedings 10th Symposium on Haptic Interfaces for Virtual Environment and Teleoperator Systems. HAPTICS 2002, pp. 354–355.
26. Ottermo, M.V., Stavdahl, O., and Johansen, T.A. (2005) Electromechanical design of a miniature tactile shape display for minimally invasive surgery. First Joint Eurohaptics Conference and Symposium on Haptic Interfaces for Virtual Environment and Teleoperator Systems (WHC'05), Pisa, Italy, 2005, pp. 561–562.
27. Ottermo, M.V., Stavdahl, O., and Johansen, T.A. (2004) Palpation instrument for augmented minimally invasive surgery. Proceedings of the IEEE/RSJ International Conference on Intelligent Robots and Systems, 2004 (IROS 2004), vol. 4, pp. 3960–3964.
28. Ikuta, K. (1990) Micro/miniature shape memory alloy actuator. Proceedings 1990 IEEE International Conference on Robotics and Automation, 1990, vol. 3, pp. 2156–2161.
29. Ikuta, K., Tsukamoto, M., and Hirose, S. (1991) Mathematical model and experimental verification of shape memory alloy for designing micro actuator. Micro Electro Mechanical Systems, 1991, MEMS '91, Proceedings. An Investigation of Micro Structures, Sensors, Actuators, Machines and Robots. IEEE, 1991, pp. 103–108.
30. Grant, D. and Hayward, V. (1997) Controller for a high strain shape memory alloy actuator: quenching of limit cycles. Proceedings 1997 IEEE International Conference on Robotics and Automation, 1997, vol. 1, pp. 254–259.
31. Wellman, P.S., Peine, W.J., Favalora, G., and Howe, R.D. (1998) Mechanical design and control of a high-bandwidth shape memory alloy tactile display. Presented at the Proceedings of the 5th International Symposium on Experimental Robotics V, 1998.
32. Velazquez, R., Pissaloux, E., Hafez, M., and Szewczyk, J. (2005) A low-cost highly-portable tactile display based on shape memory alloy micro-actuators. VECIMS 2005. Proceedings of the 2005 IEEE International Conference on Virtual Environments, Human-Computer Interfaces and Measurement Systems, 2005, 6 pp.
33. Nakatani, M., Kajimoto, H., Sekiguchi, D. *et al*. (2003) 3D form display with shape memory alloy. Proceedings International Conference on Artificial Reality and Telexistence, 2003.

34. Kontarinis, D.A., Son, J.S., Peine, W., and Howe, R.D. (1995) A tactile shape sensing and display system for teleoperated manipulation. Proceedings 1995 IEEE International Conference on Robotics and Automation, 1995, vol. 1, pp. 641–646.

35. Seung-kook, Y., Sungchul, K., Dong-Soo, K., and Hyoukreol, C. (2006) Tactile sensing to display for tangible interface. IEEE/RSJ International Conference on Intelligent Robots and Systems, 2006, pp. 3593–3598.

36. Pasquero, J. and Hayward, V. (2003) STReSS: a practical tactile display system with one millimeter spatial resolution and 700 Hz refresh rate. Eurohaptics Proceedings, Dublin, 2003, pp. 94–110.

37. Levesque, J. and Hayward, V. (2003) Experimental evidence of lateral skin strain during tactile exploration. Eurohaptics Proceedings, Dublin, 2003, pp. 261–275.

38. Hayward, V. and Cruz-Hernandez, J.M. (2000) Tactile Display Device using distributed lateral skin stretch. Proceedings of Haptic Interfaces for Virtual Environment and Teleoperator Systems Symposium, ASME/IMECE 2000, Orlando, FL, 2000, pp. 1309–1314.

39. Kyung, K., Ahn, M., Kwon, D., and Srinivasan, M.A. (2006) A compact planar distributed tactile display and effects of frequency on texture judgment. *Advanced Robotics*, **20**, 563–580.

40. Makino, Y. and Shinoda, H. (2005) Suction pressure tactile display using dual temporal stimulation modes. Presented at the SICE Annual Conference 2005, Okayama University, Okayama, Japan, 2005.

41. Makino, Y., Asamura, N., and Shinoda, H. (2004) A whole palm tactile display using suction pressure. Proceedings ICRA '04. 2004 IEEE International Conference on Robotics and Automation, 2004, vol. 2, pp. 1524–1529.

42. Makino, Y., Asamura, N., and Shinoda, H. (2004) Multi primitive tactile display based on suction pressure control. HAPTICS '04. Proceedings 12th International Symposium on Haptic Interfaces for Virtual Environment and Teleoperator Systems, 2004, pp. 90–96.

43. Rai-Choudhury, P. (ed.) (1997) *Handbook of Microlithography, Micromachining, and Microfabrication*, Micromachining and Microfabrication, Vol. **2**, SPIE Optical Engineering Press.

44. Verdu, F.V. and Gonzalez, R.N. (2003) Thermopneumatic actuator for tactile displays. Proceedings 18th Conference Design of Circuits and Integrated Systems DCIS 2003, 2003, pp. 629–633.

45. Lee, J.S. and Lucyszyn, S. (2005) A micromachined refreshable braille cell. *Journal of Microelectromechanical Systems*, **14**, 673–682.

46. Moy, G., Wagner, C., and Fearing, R.S. (2000) A compliant tactile display for teletaction. Proceedings ICRA '00. IEEE International Conference on Robotics and Automation, 2000, vol. 4, pp. 3409–3415.

47. Fearing, R.S. (1998) Teletaction System, http://robotics.eecs.berkeley.edu/~ronf/tactile.html, accessed 2012.

48. Kajimoto, H., Kawakami, N., Maeda, T., and Tachi, S. (1999) Tactile feeling display using functional electrical stimulation. Proceedings of the 9th International Conference on Artificial Reality and Telexistence, 1999.

49. Kajimoto, H., Inami, M., Kawakami, N., and Tachi, S. (2003) SmartTouch-augmentation of skin sensation with electrocutaneous display. HAPTICS 2003. Proceedings 11th Symposium on Haptic Interfaces for Virtual Environment and Teleoperator Systems, 2003, pp. 40–46.

50. Takahashi, H., Kajimoto, H., Kawakami, N., and Tachi, S. (2002) Electro tactile display with localized high-speed switching. Proceedings of the Virtual Reality Society of Japan Annual Conference, Tokyo, Japan, 2002, pp. 145–148.

51. Vos, W.K., Buma, D.G., and Veltink, P.H. (2002) Towards the optimisation of spatial electrocutaneous display parameters for sensory substitution. Proceedings 7th Annual Conference of the International Functional Electrical Stimulation Society, Ljubljana, Slovenia, 2002, pp. 332–334.

52. Hayashi, K. and Takahata, M. (2005) Objective evaluation of tactile sensation for tactile communication. *NTT DoCoMo Technical Journal*, **7**, 39–43.

53. Iwamoto, T. and Shinoda, H. (2006) High resolution tactile display using acoustic radiation pressure scanning. *Transactions of the Virtual Reality Society of Japan*, **11**, 77–86.

54. Stone, R.J. (2000) Haptic feedback: a potted History, from telepresence to virtual reality. Proceedings of the International Workshop on Haptic Human-Computer Interaction, 2000, pp. 1–7.

55. Asamura, N., Tomori, N., and Shinoda, H. (1998) A tactile feeling display based on selective stimulation to skin receptors. Proceedings of IEEE Virtual Reality Annual International Symposium, 1998, pp. 36–42.

56. Enikov, E., Lazarov, K., and Gonzales, G. (2002) Microelectrical mechanical systems actuator array for tactile communication. Proceedings of the International Conference on Computers for Handicapped Persons, Austria, 2002, pp. 551–558.

57. Nara, T., Takasaki, M., Maeda, T. *et al*. (2001) Surface Acoustic Wave (SAW) tactile display based on properties of mechanoreceptors. Proceedings IEEE Virtual Reality, 2001, pp. 13–20.
58. Maucher, T., Schemmel, J., and Meier, K. (2000) The Heidelberg tactile vision substitution system. Presented at the 6th International Conference on Tactile Aids, Hearing Aids and Cochlear Implants, ISAC2000, Exeter, 2000.
59. Konyo, M. and Todokoro, S. (2000) Artificial tactile feel display using EAP actuator. Worldwide ElectroActive Polymers, Artificial Muscles, Newsletter, Vol. 2, 2000.
60. Askins, S.A. and Book, W.J. (2003) Digital clay: user interaction model for control of a fluidically actuated haptics device. Proceedings of the 1st International Conference on Computational Methods in Fluid Power Technology, Melbourne, Australia, 2003.

8

Grayscale Graphical Softness Tactile Display

8.1 Introduction

One essential and challenging element of haptic and robotic surgery is tactile display which enables the user to experience the sense of touch by using a remote teletaction system from which information is collected and processed. Recently, different technologies have been developed to address this need and Chapters 1 and 7 present some of the state-of-the-art work in this area. This chapter elaborates on graphical softness tactile displays that can be used to convert the sense of touch into images readily recognizable by a surgeon. Using our proposed system, the surgeon can visually determine the softness of a grasped tissue and detect the presence (or absence) of unusual lumps, such as a tumor, by grasping a suspect tissue using a smart endoscopic grasper or robotic end effector equipped with appropriate tactile sensors. One major advantage of a graphical tactile display is that the surgeon receives information visually, thereby freeing his/her hands for other tasks. In the case of a physical tactile display, the display has to be in contact with the surgeon's hands which limits movement. In the case of the graphical tactile display, since there is always a monitor that the surgeon uses to view the site of operation, the tactile data can be presented to the surgeon using the same monitor. Two methods can be considered, one that allocates a corner of the monitor to show tactile information, the second superimposes the real image of the site of an operation coming from an endoscope and constructed tactile images.

This chapter is organized in two parts. The first part presents the graphical display for representing the softness of objects held by an minimally invasive surgery (MIS) grasper. The second part is devoted to detection and localizing lumps embedded inside tissue.

8.2 Graphical Softness Display

The system used for this study consists of an endoscopic grasper integrated with an array of tactile sensors, a data acquisition interface (DAQ), and appropriate signal algorithms that process information from the sensor to the display. The complete system is

Tactile Sensing and Displays: Haptic Feedback for Minimally Invasive Surgery and Robotics, First Edition.
Saeed Sokhanvar, Javad Dargahi, Siamak Najarian, and Siamak Arbatani.
© 2013 John Wiley & Sons, Ltd. Published 2013 by John Wiley & Sons, Ltd.

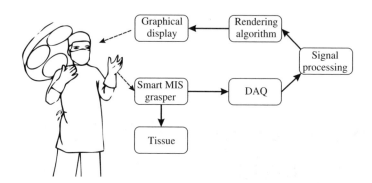

Figure 8.1 The schematic diagram of the complete system

shown schematically in Figure 8.1. When the surgeon uses the endoscopic grasper to grasp a tissue, the sensor array measures the softness of the grasped tissue under each sensing element. The electrical outputs of the piezoelectric sensing elements are then conditioned and transmitted to the DAQ system (NI PCI-6225), with which signals are amplified, filtered, digitized, and processed by a computer. A computer code was developed in the LabVIEW environment for signal conditioning such as filtering-out line noise. A representation algorithm, as later elaborated in this chapter, was used to map the features of the extracted signal to a grayscale image. Using the constructed images, the surgeon is able to establish the softness of the grasped object.

8.2.1 Feedback System

The smart endoscopic grasper consists of an array of tactile sensors, a charge amplifier, and a DAQ card. Each sensor in the tactile sensor array produces two analog voltage outputs as a result of the force applied by the grasper to the object. These voltages relate to the softness of the grasped object and interface with a DAQ card via the charge amplifiers and buffers. The connection of the sensors to the DAQ card needs extra electronic components of which further explanation is given in the following sections.

8.2.2 Sensor

The sensor unit consists of a rigid cylinder surrounded by a compliant cylinder. As shown in Figure 8.2a, a polyvinylidene fluoride (PVDF) sensing element (Goodfellow Company, USA) is positioned under both the rigid and compliant cylinders upon both of which a load is applied when an object is in contact with the sensors.

As depicted in Figure 8.2c,d, the softer the contact object is, the more of the load will transfer from the rigid cylinder to the compliant cylinder [2, 3]. The load ratio in Figure 8.2 can be calculated as follows [1, 2]:

$$\frac{F_1}{F_2} = 1 + \frac{T_2 E_1}{T_1 E_2} \tag{8.1}$$

where F_1 is the force sensed by the PVDF under the rigid cylinder, F_2 is the force sensed by the PVDF under the compliant cylinder, T_1 is the thickness of the sensed object, T_2

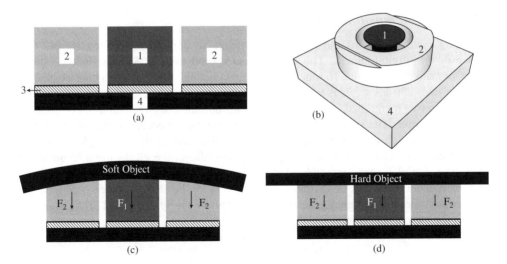

Figure 8.2   (a) Cross-sectional view of the sensing element: (1) rigid cylinder, (2) compliant cylinder, (3) PVDF, and (4) substrate. (b) Isometric view of the sensing element. (c,d) Transfer of load from rigid to compliant cylinder for a soft object and hard object, respectively [1]

is the thickness of the rigid and compliant cylinders (which are the same), E_1 is the Young's modulus of the sensed object, and E_2 is the modulus of the compliant cylinder. The larger the value of the force ratio, the stiffer the sensed object is when compared to the compliant cylinder.

The integration of eight tactile sensor units with the prototype grasper is shown in Figure 8.3.

Each sensing element in the grasper has two output signals, one from the PVDF under the rigid cylinder and the other from the PVDF under the compliant cylinder. Two arrays of four sensors are microfabricated on the grasper, one on the upper jaw and the other on the lower jaw. These two arrays, numbered from U_1 to U_4 on the upper jaw and L_1 to L_4 on the lower jaw, are shown in Figure 8.3a,b.

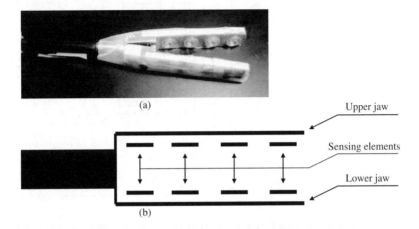

Figure 8.3   (a) Photograph of a grasper. (b) The two arrays of sensing elements

8.2.3 Data Acquisition System

The DAQ system contains the necessary hardware to convert the analog signals from the sensor into digital signals. The following DAQ :NI PCI-6225 from National Instrument Company, Resolution: 16 bits, Analog input channel: 70, Analog output channel: 4, sampling rate: 250 kS s^{-1} was used in this experiment.

As previously mentioned, each sensor has two outputs each of which is connected to a charge amplifier. The output of the charge amplifier is connected to the input channels of the DAQ through a voltage buffer, which is used to reduce the cross-talk interference between the input channels in the DAQ. Figure 8.4 shows the electronic diagram of the setup. The complete experimental setup is shown in Figure 8.5. The tactile sensor is positioned under a probe. A dynamic load is applied by the shaker, which is achieved by a power amplifier and a signal generator. The output of the sensor is fed to the connector box via the charge amplifiers.

The block diagram of the DAQ system is shown in Figure 8.6.

8.2.4 Signal Processing

The processing software, developed in the LabVIEW environment, was specifically designed for graphical demonstration of the softness of the grasped tissue. Figure 8.7 shows the steps that have been used for the construction of tactile images.

In the first step, the sensing elements on the grasper jaws touch an object. The tactile data coming from each sensing element are read in the next step. Two input channels in the DAQ are used to capture data from each sensing element. A filter is used to cancel out the 60 Hz line noise from these signals. The filtered signals are then used to calculate the

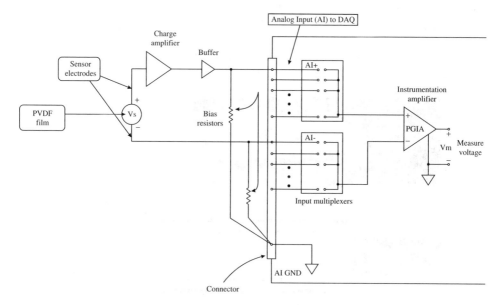

Figure 8.4 The connections from PVDF to DAQ amplifier. In this figure, only the first channel is shown. The amplifier is multiplexed between all input channels

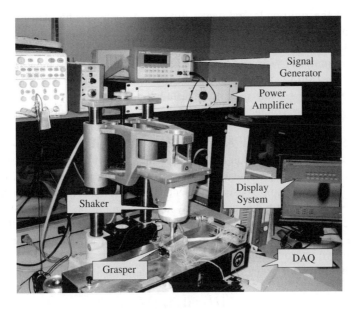

Figure 8.5   Photograph of the complete setup

Figure 8.6   Block diagram of DAQ system

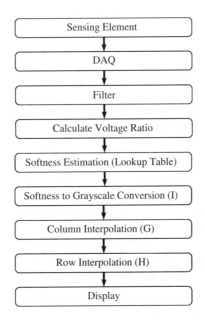

Figure 8.7   The tactile image construction flowchart

voltage ratio of the rigid and compliant cylinders in each sensing element. This voltage ratio can be used as a criterion to distinguish between different materials with varying softness [1, 3]. In order to calibrate the system, a set of reference soft objects (with known softness number) were used. In the next experiment, the standard ASTM D2240 test block kit (Instron Co.) with standard softness values (32.4, 41.9, 52.0, 61.3, 71.1, 81.1, and 88.8 Shore A) was used to measure the voltage ratios associated with each material. The resulting voltage ratio and the softness of the material were saved in a lookup table.

The voltage ratio measured for each unknown soft material was then compared to the existing reference voltage ratios on the lookup table. The softness value corresponding to the calculated voltage ratio was extracted and considered as the softness of the object which was in touch with the sensing element. This softness value was then scaled and converted into a grayscale value and displayed on the monitor. The complete image for the two tactile arrays which was derived in this way, consisted of eight cells arranged in two rows and four columns. Each cell in this 2×4 image was proportional to a sensing element in the two sensor arrays.

In step 6, the grayscale value of each cell is obtained using following relationship:

$$[I] = ([\sigma]/\alpha) K \tag{8.2}$$

where matrices $[I]$ and $[\sigma]$ represent the intensity and softness, respectively. For this study, $[I]$ and $[\sigma]$ are defined as below:

$$[I] = \begin{bmatrix} I_{U_1} & I_{U_2} & I_{U_3} & I_{U_4} \\ I_{L_1} & I_{L_2} & I_{L_3} & I_{L_4} \end{bmatrix}$$

$$[\sigma] = \begin{bmatrix} \sigma_{U_1} & \sigma_{U_2} & \sigma_{U_3} & \sigma_{U_4} \\ \sigma_{L_1} & \sigma_{L_2} & \sigma_{L_3} & \sigma_{L_4} \end{bmatrix}$$

where σ_{ij} is the softness for each sensing element obtained from the lookup table in step 5 and $[I_{ij}]$ is the grayscale value of the softness display corresponding to the same sensing element. K is a coefficient which determines the number of grayscale levels on the display. We considered 256 grayscale levels (8 bit depth) for displaying the tactile image, that is, $K = 256$. Symbol α is the softness sensitivity coefficient which enables us to show different ranges of softness on the display. We considered four different ranges for softness: Very Soft ($\alpha = 5$Shore A), Soft ($\alpha = 20$Shore A), Medium ($\alpha = 40$Shore A) and Hard ($\alpha = 100$Shore A).

Figure 8.8a shows the resulting 2×4 image when the grasper is in touch with an object. The two right-hand sensors of the grasper (U_4 and L_4) are engaged as indicated in Figure 8.8b. Furthermore, to create a smooth transition between the columns and rows, an interpolation between the grayscale values of the adjacent columns and rows is necessary. Evidently, increasing the number of cells in both directions enhances the image quality.

In step 7, a linear interpolation is performed to increase the number of columns from four to M. The resulting '$2 \times M$' matrix, $[G]$ is shown in Equation 8.3:

$$[G] = \begin{bmatrix} G_{U_1} & G_{U_2} & G_{U_3} & \cdots & G_{U_M} \\ G_{L_1} & G_{L_2} & G_{L_3} & \cdots & G_{L_M} \end{bmatrix} \tag{8.3}$$

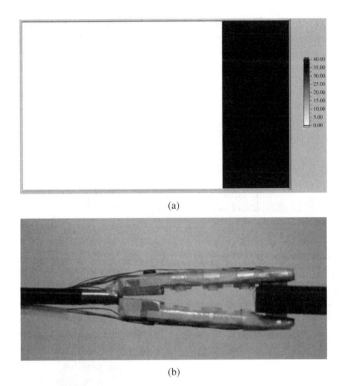

(a)

(b)

Figure 8.8 Grasper is touching an elastomeric object (with softness of 36 Shore A. (a) Softness display representation. (b) Photograph of the grasper and the grasped object

where:

$$G_{U_1} = I_{U_1}$$

$$G_{U_{\frac{M+2}{3}}} = I_{U_2}$$

$$G_{U_{\frac{2M+1}{3}}} = I_{U_3}$$

$$G_{U_M} = I_{U_4} \tag{8.4}$$

and,

$$\begin{cases} G_{U_j} = I_{U_1} + (j-1).\frac{I_{U_2}-I_{U_1}}{\frac{M-1}{3}}, & 1 < j < \frac{M+2}{3} \\ G_{U_j} = I_{U_2} + \left(j - \frac{M+2}{3}\right).\frac{I_{U_3}-I_{U_2}}{\frac{M-1}{3}}, & \frac{M+2}{3} < j < \frac{2M+1}{3} \\ G_{U_j} = I_{U_3} + \left(j - \frac{2M+1}{3}\right).\frac{I_{U_4}-I_{U_3}}{\frac{M-1}{3}}, & \frac{2M+1}{3} < j < M \end{cases} \tag{8.5}$$

G_{L_j} can be obtained in a similar way.

In the next step, another linear interpolation, as indicated in Equations 8.6 and 8.7, is implemented to increase the number of rows from 2 to N. The resulting matrix can be shown as:

$$[H] = \begin{bmatrix} H_{11} & \cdots & H_{1k} & \cdots & H_{1M} \\ H_{21} & \cdots & H_{2k} & \cdots & H_{2M} \\ H_{31} & \cdots & H_{3k} & \cdots & H_{3M} \\ & \ddots & \vdots & & \\ \vdots & & H_{ik} & & \vdots \\ & & \vdots & \ddots & \\ H_{(N-1)1} & \cdots & H_{(N-1)k} & \cdots & H_{(N-1)M} \\ H_{N1} & \cdots & H_{Nk} & \cdots & H_{NM} \end{bmatrix} \tag{8.6}$$

where:

$$H_{ik} = G_{Uk} + (i-1) \cdot \left(\frac{G_{Lk} - G_{Uk}}{N-1} \right), \quad i = 1, \ldots, N, k = 1, \ldots, M \tag{8.7}$$

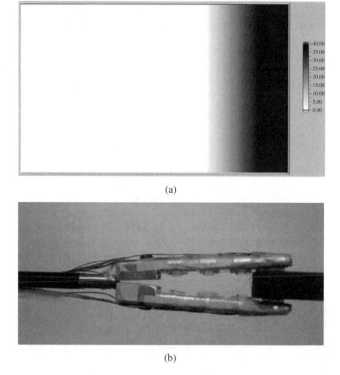

(a)

(b)

Figure 8.9   (a) The resulting tactile image after interpolation. (b) Photograph of the grasper and the grasped object

By following the above-mentioned procedures, an image is constructed based on a matrix of 60 × 100 cells, as shown in Figure 8.9.

8.2.5 Results and Discussion

The results of the conducted experiments are shown as image displays. The softness of the grasped tissue is represented by color-coding. As shown in Figure 8.10, a scale on the right-hand side of the graph shows the color coding numerically. In this case, the two right-hand sensors of the grasper (U_4 and L_4) are engaged. The softness display shows the sensed object in grayscale. Here, both U_4 and L_4 show the same softness.

This means that the grasped object has the same softness throughout its thickness. In Figure 8.11, the upper and lower jaws are grasping two different materials with different softness. The material grasped by the upper jaw is softer than the other material. The result in the softness display is presented by two different grayscales on the upper and lower part of the display.

In Figure 8.12, the two central sensors of the upper jaw, that is, U_2 and U_3, are in touch with a soft material and the other sensors on the upper and lower jaws are in touch

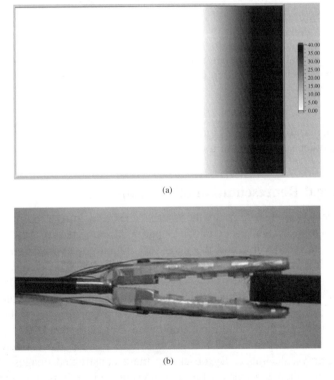

(a)

(b)

Figure 8.10 The resulting tactile image after interpolation. (b) Photograph of the grasper and the grasped object

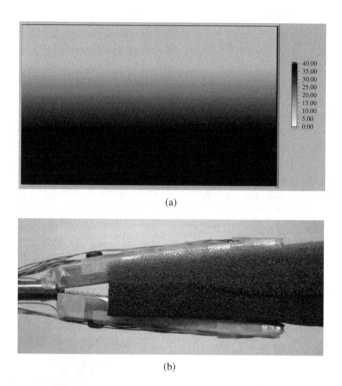

(a)

(b)

Figure 8.11   Upper and lower jaws are touching two different objects (with softnesses of 30 and 10 Shore A). The objects are located in parallel. (a) Tactile image display. (b) Photograph of the grasper and elastomers

with a harder material. Finally, in Figure 8.13, sensors U_4 and L_2 are in touch with a soft material while other sensors are in touch with a harder material.

8.3   Graphical Representation of a Lump

For the lump detection algorithm, another tactile sensor, as introduced in Chapter 4, was used. The teletaction system used for the experiment consisted of this tactile sensor, a DAQ card and its electronics, and the necessary signal-processing algorithms in order to process the information from the sensor to the display. The tactile sensor (PVDF film) measured the pressure across the seven sensing elements (1.5 mm width spaced 0.5 mm apart). Each sensor in the PVDF array produced an analog voltage output as a result of the forces applied by the object. The electrical outputs of the piezoelectric sensing elements were then conditioned and transmitted to the DAQ system. A rendering algorithm, developed in LabVIEW, was used to map the extracted features of the signal to a grayscale image. By observing these constructed images, the surgeon is able to determine whether or not a lump is present and, if it is, its approximate size and location.

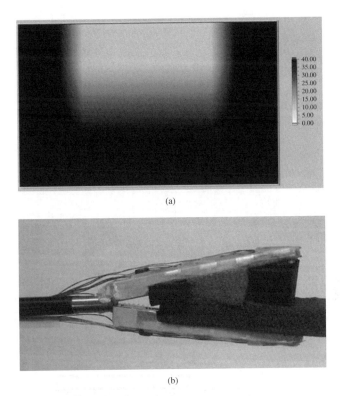

(a)

(b)

Figure 8.12 Grasper is touching two different objects (with softnesses of 30 and 10 Shore A). One embedded object is located on top. (a) Tactile image display. (b) Photograph of the grasper and elastomers

8.3.1 Sensor Structure

The details of the sensor, including its design and working concept, have already been presented in Chapter 4. The design of the deformable force-stretch array sensor is shown in Figure 8.14a. The structure of the sensor, integrated with an MIS grasper as shown in Figure 8.14b, is corrugated to ensure firm grasping of the tissue. Figure 8.14b shows the proposed grasper in which only the lower jaw is equipped with an array of tactile sensors. The sensor array consists of seven equally spaced PVDF films which are placed in parallel on an elastic material. The number of sensors, their length, width, and thickness, as well as the space between them, could be optimized for each particular application. In this study, in order to replicate the spatial resolution of a human finger, the sensor array comprised seven equally spaced piezoelectric PVDF base sensing elements.

The spatial resolution for a human finger, using the two-point discrimination threshold (TPDT), is reported to be about 2 mm [4]. Therefore, 1.5 mm wide sensing elements positioned 0.5 mm apart were considered for this study although, evidently, a finer array would yield an even better spatial resolution than that of a human finger. The PVDF can

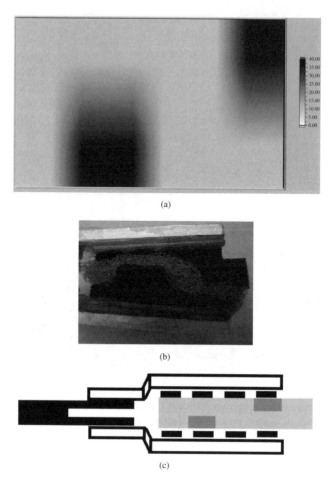

Figure 8.13 Grasper is touching two different objects (with softnesses of 30 and 10 Shore A). Two embedded objects are located on the top and bottom. (a) Tactile image display. (b) Photograph of the grasper and elastomers. (c) Grasped object with two different soften

be modeled as a voltage source with very high output impedance and, since the DAQ card requires that the input impedance be less than 100 kΩ, a buffer is necessary to match the impedance. Since the sensor consists of seven PVDF films, a total of seven analog input voltages are registered. The detailed electronic connections, from the PVDF to the differential DAQ amplifier (instrumentation amplifier), are shown in Figure 8.15.

8.3.2   Rendering Algorithm

Two configurations of the sensors on the grasper are examined. The grasper structure shown in Figure 8.14, with only one sensorized jaw, is capable of locating lumps in one dimension (x-axis), while the grasper with two sensorized jaws (which is described later in this chapter) can characterize the lumps in two dimensions, x and y axes (location of the

Figure 8.14 A view of the grasper with one active jaw equipped with an array of seven sensing elements

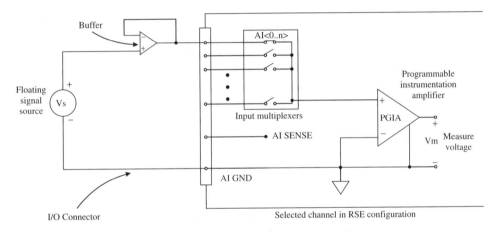

Figure 8.15 The connections from PVDF to DAQ amplifier. In this figure, only the first channel is shown. The amplifier is multiplexed between all input channels

lump in y direction can be considered as lump depth). In both designs, when there is no extraneous feature in the soft object, and depending on the grasper geometry and design, all sensing elements show either an equal output voltage or exhibit a smooth pattern that is considered to be the background frame. The presence of a lump causes an uneven voltage distribution across the sensing elements. The deduction of the background frame from the total response results in the net effect of the lump and increases the sensitivity. In addition to lump detection, the softness of the bulk soft object in the sections with no embedded masses can also be measured [5].

The outputs of the sensing elements depend on several factors, such as the ratio of the Young's modulus of the lump (E_L) to that of the tissue (E_T), the size and depth of the lump and the magnitude of the applied load. Extracting all features of the lump from the minimal sensor used in this study is a challenging task, as some combinations of lump stiffness, size, depth, and applied force create a similar output pattern. This complexity is also reported by other researchers [6]. However, there are some constraints that can be used to reduce the number of variables or, at least, to control their range. For instance, our analysis [7] shows that for $E_L/E_T > 10$, the variation of this ratio has a negligible effect on the output. On the other hand, for the absolute majority of tumors, the reported stiffness is greater than this ratio [6]. Therefore, in the practical range, the output response is not much influenced by variation in the Young's modulus of the lump. Another influential factor is the magnitude of the applied load. The contact force between the grasper and the tissue depends on the load exerted by the grasper jaws on the tissue. Therefore, it is necessary to measure the total applied load as well as the pressure distribution. The applied load can be measured in different ways. For instance, a strain gauge attached to the jaw can provide the data on the magnitude of the applied load. Another approach to measure the applied load was presented in Chapter 4 in which an extra PVDF film is used at the supports of each sensing element [8]. In the experiments conducted in this study, the load was measured using a reference load cell. Furthermore, to reduce the number of contributing parameters, we considered force as being a constant factor. The other remaining factors are the size of tumor and its location in the x and y directions. Since the majority of masses can be approximated as spherical features, the number of parameters to characterize the size of the sensor can be reduced to one value, that is, the lump radius. The first design (see Figure 8.14) overlooks the depth of lump and locates a lump merely in the x direction (the grasper length). However, by using two sets of arrays of sensing elements in the second design described in next section, it is possible to determine the depth of the lump as well.

8.3.2.1 Graphical Representation of Localized Lumps in One Dimension

As shown in Figure 8.14, in the first design the lower jaw is equipped with a sensor array, hence the upper jaw only applies compressive load to the object containing lumps. To graphically represent the location of the lump, an image with seven vertical parallel bands corresponding to the seven sensing elements was initially considered (see Figure 8.16b). The intensity of each band was considered to be proportional to the output of the corresponding sensing element. The voltage distribution along the sensor array can be considered as a vector $\{V\}_{1 \times 7}$ that is related to the intensity vector $\{I\}_{1 \times 7}$ by:

$$\begin{cases} I_i = \left(\frac{V_i}{\alpha}\right).(K-1)|V_i \leq \alpha, \quad i = 1, \dots, 7 \\ I_i = K - 1 \quad |V_i > \alpha \end{cases} \tag{8.8}$$

where α is the normalizing factor that determines the softness sensitivity (Very Soft, Soft, Medium, etc.), and K is the number of grayscales that are used to construct the graphical image (here $K = 256$). It can be seen from Equation 8.1 that for a given α, when $V_i \leq \alpha$, the scaling factor α maps the input voltage domain onto interval $[0, 1]$, then this value, using a factor of $(K - 1)$, would be mapped onto the corresponding grayscale level, between 0 and 255. Once $V_i > \alpha$, all the values of V_i would be mapped to the maximum intensity (i.e., $I_i = 255$). For example, Figure 8.16b shows the graphical display for a case where two lumps were detected in the grasped tissue. In this case, one of the lumps had been positioned above sensing element 6 and the other had been placed above and between sensing elements 2 and 3 (see Figure 8.16a for the configuration). However, due to the limited number of sensing elements, the quality of the image shown in Figure 8.16b was not satisfactory. Therefore, by using an interpolation technique, the quality of the image, as shown in Figure 8.16c, was enhanced. Prior to this interpolation, the number of elements had to be increased from 7 to any desired number (N). To do this, (N − 7) extra elements were required. Therefore, (N − 7)/6 elements were inserted between each two original elements. The resulting (1 × N) vector {G}, is in the following form:

$$\{G\} = \{G_1 \; G_2 \; \cdots \; G_{N-1} \; G_N\} \tag{8.9}$$

in which

$$G_1 = V_1, \quad G_{\frac{N+5}{6}} = V_2, \quad G_{\frac{2N+4}{6}} = V_3, \quad G_{\frac{3N+3}{6}} = V_4, \quad G_{\frac{4N+2}{6}} = V_5, \quad G_{\frac{5N+1}{6}} = V_6,$$

$$G_N = V_7$$

The intensity values assigned to the inserted elements were calculated using the linear interpolation relationship expressed in Equation 3.10:

$$G_i = V_j + \left\{ i - 1 - (j-1)\left(\frac{N+5}{6}\right)\right\} \cdot \frac{V_{j+1} - V_j}{(N-1)/6},$$
$$1 + \frac{(j-1)(N-1)}{6} < i < 1 + j\left(\frac{N-1}{6}\right) \tag{8.10}$$

where 'j' (1 ≤ j ≤ 6) and 'i' (1 ≤ i ≤ N) are indices associated with the original vector {V} and the augmented vector {G}, respectively. The numerical example for N = 60, is illustrated in Figure 8.16c.

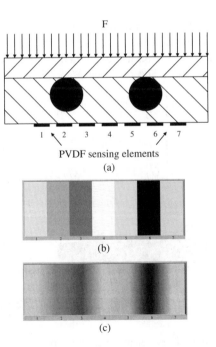

Figure 8.16 Locating the lump in one direction and its graphical rendering

8.3.2.2   Graphical Representation of Localized Lumps in Two Dimensions

Figure 8.17 illustrates the second prototype grasper in which both upper and lower jaws are equipped with arrays of sensors. Using this grasper, it is possible to locate lumps in two directions, along the jaw (x-axis) as well as its depth (y-axis).

The steps used for the construction of 2D tactile images are demonstrated in Figure 8.18. For better clarification of the algorithm used in this study, consider the case illustrated in Figure 8.19a. This figure demonstrates a grasped tissue which contains a lump that is aligned with sensing elements 2_U and 2_L, where the subscripts U and L refer to the upper and lower sensing arrays, respectively. The distance of the lump from the upper and lower sensing elements are shown by 'a' and 'b', respectively.

Figure 8.19b shows the 2D intensity graph which was built using a one-dimensional algorithm, as explained in Section 8.3.2.1. This graph consists of two rows of color bands, which correspond to two arrays of sensors, one on the top and the other at the bottom.

Therefore, this graph can be considered as a matrix with two rows (color bands) and seven columns (sensors), that is, 2×7 cells. The corresponding matrix in which each element represents voltage amplitude is in the following form:

$$[V] = \begin{bmatrix} V_{U_1} & V_{U_2} & V_{U_3} & V_{U_4} & V_{U_5} & V_{U_6} & V_{U_7} \\ V_{L_1} & V_{L_2} & V_{L_3} & V_{L_4} & V_{L_5} & V_{L_6} & V_{L_7} \end{bmatrix} \qquad (8.11)$$

As can be seen, Figure 8.19b is unable to show the precise location of the lump so, in order to provide this valuable information, the dimensions of the matrix (and consequently the number of matrix elements) were increased. The graphical enhancement in the x-direction was explained in the previous section; hence in this section the row operations (y-direction) are emphasized.

As shown in Figure 8.18, the number of rows was increased to M by inserting $(M-2)$ rows of zeros between the first and second rows of the matrix which led to an $M \times 7$ matrix. Furthermore, using the technique explained in Section 8.3.2.1, the number of columns was also increased to N. The resulting $M \times N$ matrix would be in the form:

$$[G_0] = \begin{bmatrix} G_{U_1} & G_{U_2} & \cdots & G_{U_{N-1}} & G_{U_N} \\ 0 & 0 & \cdots & 0 & 0 \\ \vdots & \vdots & \ddots & \vdots & \vdots \\ 0 & 0 & \cdots & 0 & 0 \\ G_{L_1} & G_{L_2} & \cdots & G_{L_{N-1}} & G_{L_N} \end{bmatrix} \qquad (8.12)$$

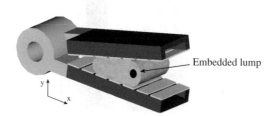

Embedded lump

Figure 8.17   The second design of grasper in which both upper and lower jaws are equipped with the sensing elements

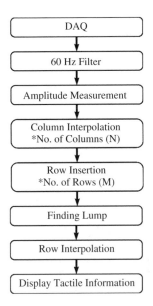

Figure 8.18   The flowchart of the algorithm implemented in LabVIEW and used for the graphical rendering

For a graphical representation of a lump, two parameters had to be determined; firstly, the location of the center of lump in each column and, secondly, its corresponding intensity value. In order to designate the vertical location of the center of the lump in each column (step 6, Figure 8.18), a relationship between the thickness of the tissue and the rows of matrix was used. If a lump is located in the tissue at a distance from the upper sensor array, it will be mapped onto row r, where r can be found from the relationship:

$$\frac{r}{M} = \frac{a}{a+b} = \frac{G_U}{G_L + G_U} \tag{8.13}$$

in which $(a + b)$ is equal to the tissue thickness and considered to be proportional to the number of rows. Regardless of whether or not a lump exists, the above equation was applied to all columns (see Figure 8.20).

If a lump exists in a column, then a and b are the distances to the center of the lump from the upper and lower sensor arrays, respectively. For the columns with no lump, the associated sensor outputs are equal and $G_L = G_U$, thus $r = M/2$. These cells are indicated in Figure 8.20 with a gray color. In other words, the algorithm assigns a nonzero value to the middle row of the columns with no lump. Although this value is not significant, it can be considered as a shortcoming of the algorithm.

In order to determine the intensity values of these locations in each column, the following relation was used:

$$G_{rj} = G_{Uj} + G_{Lj} \tag{8.14}$$

where index G_{rj} specifies the intensity value of the cell located in the row r and column j, showing the center of lump in that column. The result of this operation is a matrix, in

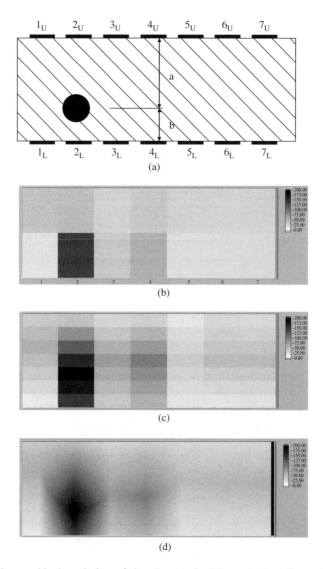

(a)

(b)

(c)

(d)

Figure 8.19 The graphical rendering of the characterized lump in two dimensions. (a) A lump located in a soft material with the upper and lower sensor arrays. (b) 2D intensity graph associated with the sensor array outputs. (c) A 7×7 matrix showing the location of the lump. (d) A 60×100 matrix that gives a better information on the location and size of the lump

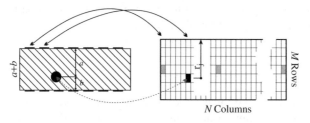

Figure 8.20 The relationship between grasped object and intensity matrix

which the centers of the detected lumps are specified:

$$[G_1] = \begin{bmatrix} G_{U_1} & G_{U_2} & \cdots & G_{U_N} \\ 0 & 0 & \cdots & 0 \\ \vdots & \vdots & \ddots & \vdots \\ 0 & 0 & \cdots & 0 \\ G_{r_1 1} & G_{r_2 2} & \cdots & G_{R_N N} \\ 0 & 0 & \cdots & 0 \\ \vdots & \vdots & \ddots & \vdots \\ 0 & 0 & \cdots & 0 \\ G_{L_1} & G_{L_2} & \cdots & G_{L_N} \end{bmatrix} \tag{8.15}$$

It should be noted that in case of multiple lumps, the center of each lump is mapped to a row that corresponds to the lump's original depth in the tissue. Therefore, for instance, $G_{r_1 1}$ and $G_{r_2 2}$ are not necessarily in the same row.

As depicted in step 7 of Figure 8.18, a row interpolation procedure was implemented. At this step, in each column, three values were known; G_{U_i}, $G_{r_i i}$, and G_{L_i}. Therefore, using these values and through a linear interpolation, a new intensity distribution was assigned to all zeros. The final intensity matrix can be represented as:

$$[H] = \begin{bmatrix} H_{11} & H_{12} & \cdots & H_{1N} \\ H_{21} & H_{22} & \cdots & H_{2N} \\ \vdots & \vdots & \ddots & \vdots \\ H_{(M-1)1} & H_{(M-1)2} & \cdots & H_{(M-1)N} \\ H_{M1} & H_{M2} & \cdots & H_{MN} \end{bmatrix} \tag{8.16}$$

where the intensity of each cell was calculated from relationship (8.17) as follows:

$$\begin{cases} H_{ij} = G_{U_j} + (i-1) \cdot \dfrac{G_{r_j j} - G_{U_j}}{r_j - 1}, |1 \leq i \leq r_j, 1 \leq j \leq N \\ H_{ij} = G_{r_j j} + (i - r_j) \cdot \dfrac{G_{L_j} - G_{r_j j}}{M - r_j}, |r_j \leq i \leq M, 1 \leq j \leq N \end{cases} \tag{8.17}$$

Figure 8.19c shows the lump position and its approximate size after implementing the mentioned algorithm when $M = N = 7$. Evidently, increasing the number of cells in both directions will enhance the quality of image. Figure 8.19d, for instance, is the constructed graphical image based on the same sensor's output and enhancement of associated matrix to $M = 60$ and $N = 100$.

8.3.3 Experiments

An experimental setup was used to generate tactile information by the application of known loads through the prototyped graspers to soft object which contained a lump. The graspers were positioned under a probe, which was equipped with a reference load cell, while the soft object and lump were sandwiched between the two jaws. Photographs of both prototype graspers, with one and two active jaws, are shown in Figure 8.21a,b, respectively. Because the PVDF base sensing elements were prepared manually, a discrepancy between the output voltages for equal loads was observed at the beginning. To compensate for this disparity, a controllable coefficient for each sensing element was defined. Then, using homogeneous elastomeric materials (without any inclusion), the outputs of the sensors were adjusted identically. The output voltages of the sensing elements in both designs were processed and displayed graphically, according to the explained algorithm.

An elastomeric material of known Young's modulus was used as the bulk soft object and three metallic balls, simulating lumps of different sizes (3.9, 6.3, and 7 mm), were inserted into hollow spaces carved out of this material. To change the depth of the lumps, several layers of the elastomeric material were cut with the same dimensions but with different thicknesses. The lumps were placed in one of the layers, so that the other elastomeric layers were used as spacers in order to increase or decrease the distance of the lump layer from the top and bottom surfaces. A similar experimental set up to that shown in Figure 8.5 was used.

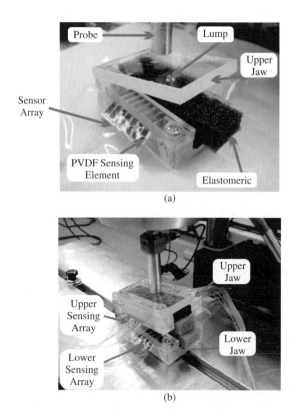

(a)

(b)

Figure 8.21 Photographs of the sensors under the test. (a) The sensor with one active jaw used for construction of one-dimension graphical images. (b) The sensor with two active jaws used for two-dimension graphical rendering of detected lumps

8.3.4 Results and Discussion

The results of the finite element analysis as well as the graphical representations of tactile information obtained from experimental cases, are shown in Figures 8.22 and 8.23 for one- and two-dimensional procedures, respectively. Each row in Figure 8.22 shows a scenario in which multiple lumps with different sizes were inserted into the elastomeric bulk material. The left column in this figure illustrates the geometrical information about the locations and size of the lumps that were placed in the soft object. The middle column in Figure 8.22 is the one-dimensional graphical representation of the sensor's outputs obtained from the experiments. The right column is the normalized voltage response of the sensing elements obtained from the finite element analysis.

In the graphical representation in Figure 8.22a (middle column), the dark column 2 has the highest intensity, showing that the lump is located above this sensing element. This can be compared with the intensity of sensing elements 4 and 5 that share a lump. For the latter elements, the maximum contact stress value occurs in a place between sensing elements 4 and 5. Therefore, each sensing element senses part of the load and, in comparison to sensing element 2, shows lower amplitude. These two elements also provide information about the size of the lump. If the middle lump was large enough to cover both sensing elements, the result would be two completely dark bands. Therefore, from the gray levels shown, the approximate size of the middle lump can be deduced. The difference observed between the outputs of sensing elements 2 and 7 can be attributed to the edge effect on the latter element. The second case (Figure 8.22b) shows two identical lumps embedded above sensing elements 2 and 5. A similar output voltage and intensity

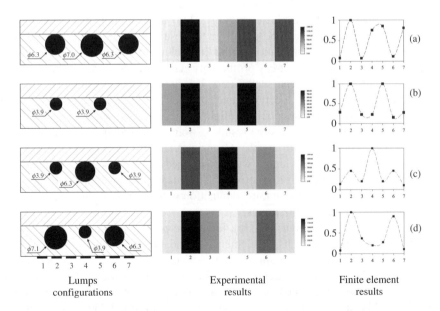

Figure 8.22 The experimental and analytical results of four case studies

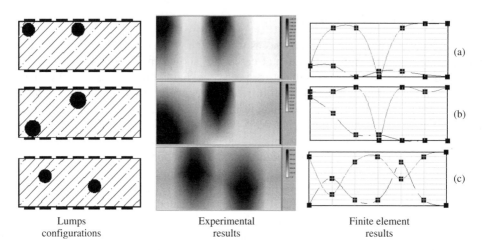

| Lumps configurations | Experimental results | Finite element results |

Figure 8.23 The experimental and analytical results for two-dimensional localization. Each row illustrates the information of the studied case. In the right-hand column, the dashed line represents the output voltages of the lower array of the sensors, while the solid line is associated with the upper jaw

can be seen in the graphical representation, as well as finite element analysis. In the third case (Figure 8.22c) a larger lump is placed between two smaller lumps. It is shown that the system is capable of detecting all three masses. However, the darker band associated with sensing element 4 gives information on the relative size of the middle lump with respect to the others. In the last case (Figure 8.22d) a small lump had been positioned between two larger lumps. As can be seen from the results, the sensor was unable to detect the smaller mass. A closer examination of the finite element stress distribution shows that the two larger lumps created a stress profile between themselves in such a way that the effect of the small mass had been suppressed. This figure demonstrates that for multiple lumps with different sizes and locations, and in order to obtain an accurate result, more than one attempt in different orientations might be needed.

Alternatively, Figure 8.23 shows the results of four case studies in which both jaws of the graspers were equipped with arrays of sensing elements. Figure 8.23a demonstrates a case in which two identical lumps were positioned close to sensing elements 1 and 4 of the upper jaw (1_U and 4_U) respectively. The corresponding graphical image shows clearly the location of the lumps and, in conjunction with the grayscale image, gives some information about the size of the lumps.

In the next configuration (Figure 8.23b), lumps are positioned apart in such a way that one lump is placed beneath sensing element 4_U and the second lump is placed above sensing element 1_L (slightly overlapped with sensing element 2_L). The graphical image constructed, based on the experimental sensor's output, is shown in the middle column of Figure 8.23b, in which the position (in x- and y-directions) is clearly extractable. The gray levels, in this case, can be compared with those of Figure 8.23c in which the lumps are positioned far from the sensing elements.

Again the position and size of lumps can be clearly perceived from the experimental data shown in the middle column. In all three cases, the finite element results shown in the right column are consistent with the experimental data. However, implementation of this algorithm may produce gray areas in the middle of the image in the absence of any lump. To rectify this problem, an enhanced algorithm is under development.

8.4 Summary and Conclusions

An MIS grasper capable of measuring the softness of the grasped objects is reported in the first part of this chapter. A feedback system is designed and incorporated into the sensor array assembly on the grasper jaws in order to transmit tactile signals from the grasper to the signal processing system. For the next step, we developed a signal processing and display system which enabled tactile data, such as tissue softness gathered from the tissue/grasper interface, to be viewed on a computer monitor, in the form of a graphical representation, and the results were discussed.

In the next part of this chapter, a system for characterizing and rendering hidden lumps in soft bulk objects is presented. The proposed system comprises an endoscopic grasper equipped with array(s) of tactile sensors, a signal processing unit, a graphical rendering algorithm, and a graphical display. This setting potentially could be used for nodule detection in laparoscopic surgery.

Initially the required information was collected from the grasped object by using a multifunctional tactile sensor that had already been developed. A unit of the utilized sensor is capable of measuring the applied contact force as well as the softness of the grasped object. An array of the sensor used in this study is also capable of reporting the location and size of the lump. The output voltages of the sensing elements are buffered, digitized, filtered, and transmitted to a computer. Then, through a rendering algorithm developed in the LabVIEW environment, the tactile data were transformed to grayscale images and were displayed on a monitor.

The experiments on the prototype graspers were conducted and the data were graphically rendered on a display. Hard objects were inserted into pre-determined positions in elastomeric material and were held by the grasper and the calculated images were compared with known images.

Graphical rendering of localized objects is a feasible technique with great potential for use in MIS. Using this method, a part of the lost tactile information available from palpation can be restored. This capability is useful not only for MIS, but also for MIS robotic surgery and robotic surgery in general. This research also paves the way for other anatomical applications elsewhere, such as beating arteries that could potentially be detected and graphically rendered.

References

1. Narayanan, N.B., Bonakdar, A., Dargahi, J. *et al.* (2006) Design and analysis of a micromachined piezoelectric sensor for measuring the viscoelastic properties of tissues in minimally invasive surgery. *Smart Materials and Structures*, **15**, 1684–1690.
2. Dargahi, J., Najarian, S., and Zheng, X.Z. (2005) Measurements and modeling of compliance using novel multi-sensor endoscopic grasper. *Sensors and Materials*, **17**, 7–20.
3. Dargahi, J. (2002) An endoscopic and robotic tooth-like compliance and roughness tactile sensor. *Journal of Mechanical Engineering Design (ASME)*, **124**, 576–582.

4. Geiger, S.R. (ed.) (1984) Section 1: the nervous system, *Handbook of Physiology*, American Physiological Society.
5. Wellman, P.S., Dalton, E.P., Krag, D. *et al*. (2001) Tactile imaging of breast masses: first clinical report. *Archives of Surgery*, **136**, 204–208.
6. Wellman, P.S. and Howe, R.D. (1999) Extracting features from tactile maps. Proceedings of the MICCAI, pp. 1133–1142.
7. Sokhanvar, S., Dargahi, J., and Packirisamy, M. (2008) Hyperelastic modelling and parametric study of soft tissue embedded lump for MIS applications. *International Journal of Medical Robotics and Computer Assisted Surgery*, **4**, 232–241.
8. Sokhanvar, S., Packirisamy, M., and Dargahi, J. (2004) A novel PVDF based softness and pulse sensor for minimally invasive surgery. Presented at the 3rd IEEE International Conference on Sensors, Austria.

9

Minimally Invasive Robotic Surgery

In minimally invasive surgery (MIS), operations are carried out by surgeons using slender instruments and viewing equipment inserted into the body through small incisions. MIS has become an established practice over the years and, for certain surgical procedures, has become *de rigueur*. In recent years, however, minimally invasive robotic surgery (MIRS) is making inroads into this practice with ever-increasing frequency. This revolutionary approach to surgery has been made possible by recent radical improvements in robotic methods and techniques. For the surgeon, this computer-enhanced technology and robotic precision ensures a level of surgical precision that was never previously possible. The use of robotics is changing medicine dramatically. As technology continues to advance and patients experience the benefits of robotic surgery, the demand for robotic procedures will continue to increase.

The most significant advantage of robotic surgery, certainly as far as the patient is concerned, is reduced pain and scarring, since, by using cameras and enhanced visual effects, doctors can make the smallest of incisions. The da Vinci™ and ZEUS™ systems both use 'arms' to operate and, in order for these arms to get inside the body and operate, they only need an incision of a few centimeters. It is reported that some cardiac bypass surgeries were successfully done using only three incisions, each about 1 cm in length. Typically, for that type of surgery, an incision of about 30 cm in length is required. It is this and its many other advantages that justifies the use of MIRS, notwithstanding the fact that it is still in the process of being fully tried and tested. Due to the small and precise incisions, the patient's hospital stay is greatly reduced and the patient in the case just mentioned required only 12 hours recovery period before being discharged. Typically, a person needs far less recovery time from a 3 cm scar rather than one which is 10 times larger. Also, the risk of infection or complications decreases commensurately with incision size.

Besides the obvious rewards to the patient, robotic surgery is also very advantageous to the surgeon and the hospital itself. In the ZEUS™ Surgical System, an 'arm' on the machine is dedicated to the automated endoscopic system for optimal positioning (AESOP) which is a 3D camera used in robotic surgery. It can be zoomed in by either voice activation or pedals located at the surgeon's feet. Doctors who have used this actually

Tactile Sensing and Displays: Haptic Feedback for Minimally Invasive Surgery and Robotics, First Edition.
Saeed Sokhanvar, Javad Dargahi, Siamak Najarian, and Siamak Arbatani.
© 2013 John Wiley & Sons, Ltd. Published 2013 by John Wiley & Sons, Ltd.

argue that AESOP gives a better image than real life. This is particularly true with surgeons that have less than perfect vision or for those who are performing microscopic surgery that deals with nerves. Also, by using hand controls, the surgeons can reach places in the body that are normally unreachable by the human hand. Finally, one of the more evident advantages of robotic surgery, is in long operations, particularly those that deal with nerve or tissue reconstruction. Surgeons often become tired after performing microscopic surgical procedures that last many hours. However, by having the ability to be seated and have less strain on the eyes, doctors can control their natural flinching or nerves more efficiently. So far, robots have been used to position an endoscope, perform gallbladder surgery and correct gastroesophageal reflux and heartburn. A current ambitious goal within the robotic surgery field is to design a robot that can be used to perform closed-chest, beating-heart surgery. The current growth rate of this technology is an indication that the use of robotics in surgery will increase significantly over the next decades.

In heart surgery the introduction of endoscopic techniques was promising, but not as satisfying as the application of robots in other surgical disciplines [1–3]. Thus far, the results of complex cardiac surgery are less than satisfactory due to tremor in the long instruments and inadequate freedom of movement. Heart surgery, using pure endoscopic techniques, has still not yet been fully established, since high-precision specialty instruments are lacking in this field. The purpose of telemanipulated endoscopic assistance is to eliminate many of these impediments and, in this regard, enhancements have been made gradually with regard to motion scaling, tremor filtration, three-dimensional vision systems, and improving the fulcrum effect. Implementation of these advances would enable the surgeon to operate using such a surgical mechatronic system in a more comfortable, dextrous, and intuitive manner. Part of the solution is to implement telemanipulators that would provide more than six degrees of freedom in movement (rather than the four that is found in conventional endoscopic instruments). By virtue of this, the surgeon will have as many degrees of freedom in movement, as in conventional open surgery. Furthermore, this telemanipulator system would be remotely controlled by the surgeon, provide 3D-optic viewing capabilities and be tremor-free [4]. The culmination of all this, 10 years later, was the implementation of totally endoscopic heart surgery using the telemanipulator da Vinci™ system (Intuitive Surgical, Inc., Sunnyvale, CA, USA) which was originally introduced for endoscopic abdominal surgery.

However, for patients who require valve surgery, congenital heart surgery, and bypass surgery, only a minority are able to avail themselves of telemanipulated technology due to certain inequities in the system. One significant limitation is the necessity of acquiring haptic feedback, a matter which is still a matter of controversy for both robotically working surgeons and haptic engineers [5–7] and is the focus of different chapters in this book.

For both the virtual and artificial scenarios, tactile sensing and haptic feedback is an essential parameter [8, 9] and is still the subject of much discussion, due its importance in the field of surgical telepresence of remote objects. Although micro-surgical telerobotic systems have been addressed by many research groups [10–12], many important questions and problems remain unanswered, or at least have not been answered or solved sufficiently, in the area of haptic feedback. Another basic and unsolved problem with telemanipulated surgery, and for which further exploratory work is required, is the sudden breaking of surgical suture material with resulting tissue damage. It is certainly evident that the ability to provide haptic feedback during robotic surgery would greatly alleviate the fatigue that is placed on the surgeon, notwithstanding the enhanced visual skill and movements that this procedure otherwise provides [13].

9.1 Robotic System for Endoscopic Heart Surgery

Because robotic devices are able to perform repeated and routine manipulation tasks with greater dexterity and frequency than a human, their place in the medical and surgical markets is all but assured. Some of the weaknesses in current robotic devices, such as substantial lack of haptic feedback and adaptability, will be discussed. At the present time it is not possible to 'program' a robot to perform steps of a surgical operation autonomously. Nevertheless, these limitations do not prevent robots from being useful in the operating room, although extensive human, technical and surgical input, guidance, and intervention are still required. Surgical robots can be viewed as 'extending' and 'enhancing' human capabilities rather than replacing surgeons, in contrast to the example of industrial replacement of humans by robots.

Intuitive Surgical™ intended to create, in conjunction with the da Vinci™ Surgical System, a conception of a surgeon–robot interface so transparent to the surgeon that his/her set of skills could be used in a natural and instinctive manner. Its accurate visualization is critical since visual cues are used to compensate for the loss of haptic feedback. Haptic feedback is currently limited to interaction with rigid structures, such as tool-on-tool interfaces, not soft tissue, which requires the surgeon to rely on visual feedback in tasks such as suturing. Research groups began to analyze the use of haptic feedback when using fine suture material [14, 15], but their findings were of little or no interest to the medical fraternity, including heart surgeons. The basic consideration is to offer the heart surgeon an additional sensory channel, in addition to the visual channel, not only to avoid breaking surgical suture material with ensuing tissue damage, but also to decrease visual fatigue. In this regard, 17 new applications of this technology are beginning to emerge in conjunction with feedback from creative surgeons [16, 17]. Nevertheless, present-day robotic surgical systems have limitations that have slowed its widespread introduction and continuation. One major problem, as previously mentioned, is the lack of haptic feedback [18, 19]. A second major concern is the cumbersome and none-too versatile nature of the robotic system itself, so it is quite easy to envision integrated imaging, navigation, and enhanced sensory capabilities being available in the next generation of telesurgical systems [20, 21]. Braun et al. [22] conducted experiments to examine claims about the necessity of force feedback for robot-assisted surgical procedures in cardiac surgery. They presented a novel approach of a robotic system for minimally invasive and endoscopic surgery, with the main purposes of evaluating force feedback and machine learning. However, in their experimental setup, although the feel for different and radically changed tissues could not be analyzed in sufficient depth, their experiments did, nonetheless, show that haptic feedback can be employed to prevent the surgeon from making potentially harmful mistakes. Tension of thread material and tissue parts can be measured and displayed in order to restrict force application to tolerable amplitudes. In addition, the collision of instruments can be detected and intercepted by the evaluation of real-time forces. In their proposed system, forces are measured at the surgical instruments using multi-dimensional haptic styluses, and fed back to the surgeon's hands. They also incorporated the results of their experiments into control software, for modeling and simulation of haptic interaction with a tissue model, and demonstrated that the surgical procedure in robotic heart surgery is safer, quicker, and gentler for the patient and more comfortable for the surgeon when using force feedback.

9.2   da Vinci™ and Amadeus Composer™ Robot Surgical System

The da Vinci™ Surgical System is a robotic surgical system manufactured by Intuitive Surgical and designed to facilitate complex surgery using a minimally invasive approach. The system is controlled by a surgeon from a console. It is commonly used for prostatectomies (surgical removal of part of the prostate gland), and increasingly for cardiac valve repair and gynecologic surgical procedures [23].

The da Vinci™ System consists of a surgeon's console, that is typically in the same room as the patient, and a patient-side cart with four interactive robotic arms controlled from the console. Two of the arms hold tools such as scissors and unipolar or bipolar electrocautery instruments, the third acts as a scalpel. The fourth arm holds an endoscopic camera having two lenses that gives the surgeon full stereoscopic vision from the console. The surgeon sits at the console and looks through two eye holes at a 3-D image of the procedure, meanwhile maneuvering the arms with two foot pedals and two hand controllers. The da Vinci™ System scales, filters, and translates the surgeon's hand movements (Figure 9.1a) into more precise micro-movements of the instruments (Figure 9.1b,c,d,e), which operate through small incisions in the body.

To perform a procedure, the surgeon uses the console's master controls to maneuver the patient-side cart's three or four robotic arms (depending on the system model), which secures the instruments and a high-resolution endoscopic camera. The instruments' jointed-wrist design exceeds the natural range of motion of the human hand and motion scaling and tremor reduction further interpret and refine the surgeon's hand movements. The da Vinci™ System incorporates multiple, redundant safety features designed to minimize opportunities for human error when compared with traditional approaches. At no time is the surgical robot in control or autonomous; it operates on a 'master/slave' relationship, the surgeon being the 'master' and the robot being the 'slave.'

The da Vinci™ System has been designed to improve upon conventional laparoscopy, in which the surgeon operates while standing, using hand-held, long-shafted instruments, which have no wrists. With conventional laparoscopy, the surgeon must look up and away from the instruments, to a nearby 2D video monitor to see an image of the target anatomy. The surgeon must also rely on his/her patient-side assistant to position the camera correctly. In contrast, the da Vinci™ System's ergonomic design allows the surgeon to operate from a seated position at the console, with eyes and hands positioned in line with the instruments. To move the instruments or to reposition the camera, the surgeon simply moves his/her hands.

By providing surgeons with superior visualization, enhanced dexterity, greater precision, and ergonomic comfort, the da Vinci™ Surgical System makes it possible for more surgeons to perform minimally invasive procedures involving complex dissection or reconstruction. For the patient, a da Vinci™ procedure can offer all the benefits of a minimally invasive surgical procedure with less pain and blood loss, and hence a reduced need for blood transfusions. Moreover, the da Vinci™ System enables a shorter hospital internment, a speedier recovery, and quicker return to normal daily activities [24].

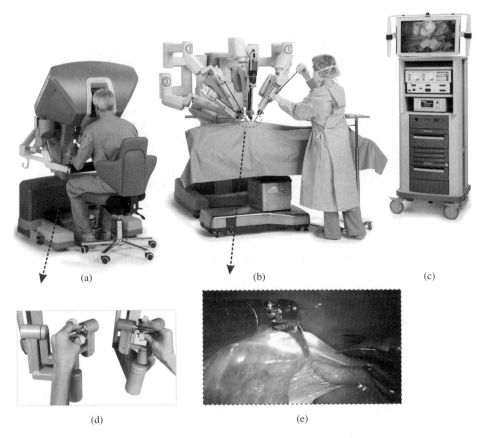

Figure 9.1 da Vinci™ (a) surgeon console, (b) patient cart, (c) vision cart, (d) master controls, and (e) robotic arms equipped with surgical instruments (© 2012 Intuitive Surgical, Inc.) (See Plate 17)

The Amadeus Composer™ Robotic Surgical System (Figure 9.2) from Titan Medical Inc. is planned for animal and tissue feasibility studies, and human clinical trials in 2012–2013, and FDA 510(K) submission in the second half of 2014. Compared with the da Vinci™ system, Amadeus Composer™ is a four-armed robotic surgical platform with much smaller and sleeker external snake-like robotic arms (made by the Kuka Robot Group) tailored for more dexterity inside the body, which enables substantially more procedures to be performed compared to other robot-assisted surgical equipment. The enhanced vision system (1080p 3D HD vision) of Amadeus™ provides the surgeon with a clearer image of anatomical details. The open architecture provides increased flexibility for expansion of robotic surgery into untapped specialties and procedures. This robotic surgical system is equipped with a force feedback mechanism, for the first time, which enables feeling of tissue forces and emulates the natural touch. Force feedback in this system lets the surgeons know how hard they are pulling on tissues, sutures, or other biological structures, in order to prevent tissue damage [25, 26].

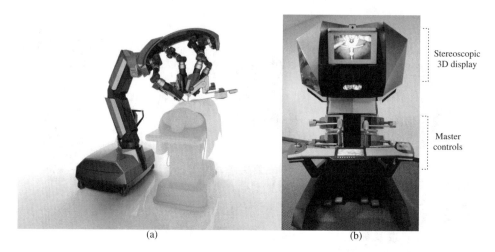

Figure 9.2   Amadeus Composer™ Robotic Surgical System, (a) Patient cart and robotic arms, (b) Master station (© 2012 Titan Medical, Inc.)

9.3   Advantages and Disadvantages of Robotic Surgery

The advantages of robotic surgery systems are numerous, because they overcome many of the obstacles of laparoscopic surgery. They increase dexterity, restore proper hand/eye harmony and an ergonomic position, and improve visualization. In addition, these systems now enable surgeries that were previously either technically difficult or even unfeasible. The advantages and disadvantages of conventional surgery, MIS and MIRS, on a comparative basis, are listed in Table 9.1.

These robotic systems enhance dexterity in several ways. Firstly, they are equipped with instruments that provide increased degrees of freedom, which greatly enhance the surgeon's ability to manipulate instruments that grasp the tissues. Secondly, they are designed in such a way that any hand tremor can be compensated in the end-effector motion through appropriate hardware and software filters. Thirdly, they scale movements so that large movements of the control grips can be transformed into micro-motions inside the patient [27].

Another important advantage is the restoration of proper hand/eye coordination and an ergonomic position. These robotic systems eliminate the fulcrum effect, making instrument manipulation more intuitive. With the surgeon sitting at a remote, ergonomically designed workstation, current systems also eliminate the need to twist and turn in awkward positions to move the instruments, while at the same time viewing the monitor.

By most accounts, the enhanced vision afforded by these systems is remarkable. The three-dimensional view with its incumbent depth perception is a marked improvement over the views supplied by conventional laparoscopic cameras. The surgeon also has the ability to control a stable visual field directly, with increased magnification and maneuverability. All of this creates images with increased resolution that, combined with the

Table 9.1 Advantages and disadvantages of conventional surgery, MIS and MIRS

	Advantages	Disadvantages/Limitations
Conventional surgery	Strong hand/eye harmony	Limited dexterity outside natural scale
	Dexterous	Prone to tremor and fatigue
	Flexible and adaptable	Limited geometric accuracy
	Can integrate extensive and diverse information	Limited ability to use quantitative information
	Rudimentary haptic abilities	Limited sterility
	Able to use qualitative information	Susceptible to radiation and infection
	Good judgment	
	Easy to instruct and debrief	
MIS	Well-developed technology	Loss of touch sensation
	Affordable and ubiquitous	Loss of 3D visualization
	Proven efficacy	Compromised dexterity
		Limited degrees of motion
		The fulcrum effect
		Amplification of physiologic tremors
MIRS	3D visualization	Absence of haptic feedback
	Improved dexterity	Very expensive
	Higher degrees of freedom	High start-up cost
	Elimination of fulcrum effect	May require extra staff to operate
	Elimination of physiologic tremors	No judgment
	Ability to scale motions	Unable to use qualitative information
	Micro-anastomoses possible	Expensive
	Tele-surgery	Technology in flux
	Ergonomic position	More studies needed
	Good geometric accuracy	
	Stable and untiring	
	Can use diverse sensors in control	
	May be sterilized	
	Resistant to radiation and infection	

increased degrees of freedom and enhanced dexterity, greatly enhances the surgeon's ability to identify and dissect anatomic structures, as well as being able to undertake micro-anastomoses. Other major advantages of robotic surgery are decreased blood loss, less pain, quicker healing time, and less use of pain medication [28].

There are, however, several disadvantages to robotic surgical systems. Firstly, robotic surgery is a new technology and its uses and efficacy have not yet been well established. To date, mostly studies of feasibility have been conducted, with almost no long-term follow-up studies. Many procedures also have to be redesigned to optimize the use of robotic arms and increase efficiency, although it is fully hoped and supposed that these disadvantages will be remedied with the advent of time.

Secondly, the cost of robotic systems is prohibitive for many institutions [29]. The extent to which these costs become less with the passage of time is dependant largely on improvements in technology and the extent, therefore, to which this technology becomes affordable and hence more widely accepted [27]. Others believe that the incumbent investments required to improve haptics technology, together with increased processor speeds and more complex software, will increase the cost of these systems [30]. Also at issue is the problem of upgrading systems which will pose the question to hospitals and healthcare organizations as to how much they will need to spend on upgrades and how often. In any case, many believe that to justify the purchase of these systems they must first gain widespread and multi-disciplinary usage [30].

Thirdly, these systems are bulky and are equipped with cumbersome robotic arms which, in today's already crowded operating rooms, could cause space issues with the surgical team [30]. Given the fact that it is difficult for both the surgical team and the robot to fit into the operating room, and that expanding the size of the surgical environment is hardly practical and certainly not cost-effective, the only practical alternative is to further miniaturize the robotic arms and instruments.

One of the potential disadvantages identified is a lack of compatible instruments and equipment, which necessitates the inclusion of table-side assistants to perform part of the surgery [27]. This, however, is a transient disadvantage because new technologies have, and will, be developed to address these shortcomings. Most of these disadvantages will be remedied with improvements in technology and, only with time will it be known if the use of these systems justifies their cost. Certainly, if the cost of these systems continues to remain high, and their inclusion in routine surgeries does not reduce the cost, it is unlikely that there will be a robot in every operating room.

9.4 Applications

Several robotic systems are currently approved by the FDA for specific surgical procedures. ROBODOC™ is used to precisely core out the femur in hip replacement surgery. Computer Motion Inc. of Goleta, CA, has two systems on the market. One, called AESOP, is a voice-controlled endoscope with seven degrees of freedom. This system can be used in any laparoscopic procedure to enhance the surgeon's ability to control a stable image. The ZEUS™ and da Vinci™ systems have been used in a variety of disciplines for laparoscopic surgeries, including cholecystectomies, mitral valve repairs, radical prostatectomies, reversal of tubal ligations, in addition to many gastrointestinal surgeries, nephrectomies, and kidney transplants. The number and types of surgeries being performed with robots is increasing rapidly as more institutions acquire these systems, one of their more notable uses being in totally endoscopic coronary artery grafting, a procedure formerly outside the limitations of laparoscopic technology.

The amount of data being generated on robotic surgery is growing rapidly, and the early data are promising. Many studies have evaluated the feasibility of robot-assisted surgery. One study by Cadiere *et al*. [31] evaluated the feasibility of robotic laparoscopic surgery on 146 patients. Procedures performed with a da Vinci™ robot included

39 anti-reflux procedures, 48 cholecystectomies, 28 tubal reanastomoses, 10 gastroplasties for obesity, 3 inguinal hernia repairs, 3 intrarectal procedures, 2 hysterectomies, 2 cardiac procedures, 2 prostatectomies, 2 artiovenous fistulas, 1 lumbar sympathectomy, 1 appendectomy, 1 laryngeal exploration, 1 varicocele ligation, 1 endometriosis cure, and 1 neosalpingostomy. This study found robotic laparoscopic surgery not only to be feasible, but also found the robot to be most useful in intra-abdominal micro-surgery or for manipulations in very small spaces, and reported no robot-related morbidity. Another study by Falcone *et al*. [32] tested the feasibility of robot-assisted laparoscopic micro-surgical tubal anastomosis. In this study, 10 patients who had previously undergone tubal sterilization underwent tubal reanastomosis. They found that the 19 tubes were reanastomosed successfully, of which 17 were still patent six weeks postoperatively. There have been five pregnancies in this group so far. Margossian and Falcone [33] also studied the feasibility of robotic surgery in complex gynecologic surgeries in pigs. In this study, 10 pigs underwent adnexal surgery or hysterectomy using the ZEUS™ robotic system. They found that robotic surgery is safe and feasible for complex gynecological surgeries. In yet another study by Marescaux *et al*. [34], the safety and feasibility of telerobotic laparoscopic cholecystectomy was tested in a prospective study of 25 patients undergoing the procedure, of which 24 were performed successfully, and one was converted to a traditional laparoscopic procedure. This study concluded that robotic laparoscopic cholecystectomy is safe and feasible. Another study, by Abbou *et al*. [35], found telerobotic laparoscopic radical prostatectomy to be feasible and safe, with dramatically enhanced dexterity.

One of the areas where robotic surgery is transforming medicine the most, and one of the areas generating the most excitement, is in minimally invasive cardiac surgery. Several groups have developed robotic procedures that expand laparoscopic techniques into this previously unexplored territory, with encouraging results. Prasad *et al*. successfully constructed left internal thoracic artery (LITA) to left anterior descending (LAD) artery anastomoses on 17 of 19 patients with the use of a robotic system [32]. They concluded that robotically assisted endoscopic coronary bypass surgery showed favorable short-term outcomes with no adverse events and found robotic assistance to be an enabling technology that allows surgeons to perform endoscopic coronary anastomoses. Damiano *et al*. [36] conducted a multi-center clinical trial of robotically assisted coronary artery bypass grafting. In this study 32 patients scheduled for primary coronary surgery underwent endoscopic anastomosis of the LITA to LAD. A two-month follow-up revealed a graft patency of 93%, after which it was concluded that robot-assisted coronary bypass grafting is feasible. In another study, Mohr *et al*. [37] used the da Vinci™ system to perform coronary artery bypass grafting on 131 patients and mitral valve repair on 17 patients. They used the robot to perform LITA take down, LITA-LAD anastomosis in standard sternotomy bypass, and total endoscopic coronary artery bypass grafting LITA-LAD anastomosis on the arrested heart and the beating heart. They found that robotic systems could be used safely with selected patients to perform endoscopic cardiac surgery. Internal thoracic artery takedown is an effective modality, and total endoscopic bypass on an arrested heart is feasible, but does not offer a major benefit to the minimally invasive direct approach, because cardiopulmonary bypass is still required. Their study suggests that robotic systems have not yet advanced far enough to perform endoscopic closed-chest beating-heart bypass grafting, despite some technical success in two of eight patients. In addition, robotic endoscopic mitral valve repair was successful in 14 of 17 patients. By contrast, several groups in Europe have successfully performed closed-chest, off-pump coronary artery bypass grafting using an endoscopic stabilizer. Kappert *et al*. [38] performed 37 off-pump totally endoscopic coronary artery bypasses (TECABs) on a beating heart using the da Vinci™ system and an endoscopic stabilizer. In this series, they reported a 3.4% rate of conversion to median sternotomy. They concluded that their results promote optimism about further development of TECAB. Another study by Boehm *et al*. [39] using a similar stabilizer to the ZEUS™ system had similar results and conclusions about TECAB. Interestingly, a study by Cisowski and Drzewiecki in Poland [40] compared percutaneous stenting with endoscopic coronary artery bypass grafting in patients with single-vessel disease. In this series of 100 patients, percutaneous stenting resulted in restenosis in 6 and 12% at 1 and 6 months, respectively, compared with 2% at 6 months in the endoscopic bypass group.

Another use for robotic systems that is being investigated is pediatric laparoscopic surgery. Currently, this type of surgery is limited by an inability to perform precise anastomoses of 2–15 mm [41]. Although laparoscopic techniques may be used to treat infants with intestinal atresia, choledochal cysts, biliary atresia, and esophageal atresia, it is not the standard approach because of technical difficulties. To evaluate the feasibility of robotic systems in pediatric minimally invasive surgery, Hollands and Dixey [42] developed a study where enteroenterostomy, hepaticojejunostomy, and portoentorostomy were performed on piglets. They found all the procedures to be technically feasible with the ZEUS™ robotic system. The study concludes that robot-assisted laparoscopic techniques are technically feasible in pediatric surgery and may be of benefit in treating various disorders in term and preterm infants. More recently, Hollands and Dixey [43] devised a study using 10 piglets to develop the procedure and evaluate the feasibility of performing a robot-assisted esophagostomy. In this study, robot-assisted and thoracoscopic approaches were evaluated and compared for leak, narrowing, caliber, mucosal approximation, as well as anesthesia, operative, anastomotic, and robotic set-up times and found that, in all, the robotic-assisted approach is feasible. They also discerned no statistically significant difference between the two approaches based on the above variables. Notwithstanding the feasibility of robotic surgical systems, further high-quality clinical trials still need to be performed before their full potential can be realized.

9.4.1   Practical Applications of Robotic Surgery Today

In today's competitive healthcare market, many organizations are interested in making themselves 'cutting-edge' institutions with the most advanced technological equipment and the very newest treatment and testing modalities and, by so doing, allow themselves to capture more of the healthcare market. Acquiring a surgical robot is, essentially, the entry fee into marketing an institution's surgical specialties as 'the most advanced.' It is not uncommon, for example, to see a photograph of a surgical robot on the cover of a hospital's marketing brochure and yet see nothing within it that mentions robotic surgery.

Although surgical robotics is still in its embryonic stage there is, nonetheless, the full expectation that rapid and strident technological advances will secure its place in the appliances of tomorrow. Already, the development of robotics is spurring interest in new tissue anastomosis techniques, improving laparoscopic instruments, and digital integration of already existing technologies.

As previously mentioned, applications of robotic surgery are expanding rapidly into many different surgical disciplines. Although the costs of these systems are high, their contribution to the enhancement of medical practices is undoubted and their presence in any medical institution is likely to become more inevitable with the advent of time, although this still remains to be seen.

9.5   The Future of Robotic Surgery

There will be many questions regarding the use of robotics surgery, including malpractice suits, liability, credentialing, training requirements, and interstate licensing for telesurgeons. However, these obstacles will undoubtedly be resolved with over time and the hitherto proven advantageous factors of robot-assisted surgery will ensure its continued development and expansion. For example, the sophisticated controls and the multiple degrees of freedom afforded by the ZEUS™ and da Vinci™ systems allow increased mobility and no tremor without comprising the visual field, to make micro-anastomosis possible. Many have made the observation that robotic systems are information systems and, as such, they have the ability to interface and integrate many of the technologies being developed for and currently used in the operating room [30]. One exciting possibility is expanding the use of preoperative (computed tomography or magnetic resonance) and intraoperative video image fusion to better guide the surgeon in dissection and identifying pathology. These data may also be used to rehearse complex procedures before they are undertaken. The nature of robotic systems also makes the possibility of long-distance intraoperative consultation or guidance possible and it may provide new opportunities for teaching and assessment of new surgeons through mentoring and simulation. Computer Motion, the manufacturers of the ZEUS™ robotic surgical system, is already marketing a device called SOCRATES™ that allows surgeons at remote sites to connect to an operating room and share video and audio, to use a 'telestrator' to highlight anatomy, and to control the AESOP endoscopic camera.

Technically, much remains to be done before the full potential of robotic surgery can be realized. Although these systems have greatly improved in dexterity, they have yet to develop the full potential in instrumentation or to incorporate the full range of sensory input, and it is evident that more standard mechanical and energy-directed tools need to be developed. Some authors also believe that robotic surgery can be extended into the realm of advanced diagnostic testing, with the development and use of ultrasonography, near infrared, and confocal microscopy equipment [44].

As much as robots are conjured in popular culture and literature, the future of robotics in surgery is limited only by the imagination. Many future 'advancements' are already the subject of ongoing research. Some laboratories, including the authors' laboratory, are currently working on systems to relay touch sensation from robotic instruments back to the surgeon [21, 45–49]. Other laboratories are working on improving current methods and developing new devices for anastomoses that require no sutures [41, 50, 51]. When most people think about robotics, they think about automation. The possibility of automating some tasks is both exciting and controversial. Future systems, which are limited only by the imagination, might include the ability for a surgeon to program the surgical procedure itself and then to merely adopt a supervisory role as the robot performs most of the tasks autonomously.

References

1. Bholat, O.S., Haluck, R.S., Murray, W.B. *et al*. (1999) Tactile feedback is present during minimally invasive surgery. *Journal of the American College of Surgeons*, **189**, 349–355.
2. Gutt, C.N., Oniu, T., Mehrabi, A. *et al*. (2004) Robot-assisted abdominal surgery. *British Journal of Surgery*, **91**, 1390–1397.
3. Mitsuishi, M., Tomisaki, S., Yoshidome, T. *et al*. (2000) Tele-micro-surgery system with intelligent user interface. Proceedings of the ICRA '00. IEEE International Conference on Robotics and Automation, 2000, vol. 2, pp. 1607–1614.
4. Suematsu, Y. and del Nido, P.J. (2004) Robotic pediatric cardiac surgery: present and future perspectives. *The American Journal of Surgery*, **188**, 98–103.
5. Bethea, B.T., Okamura, A.M., Kitagawa, M. *et al*. (2004) Application of haptic feedback to robotic surgery. *Journal of Laparoendoscopic & Advanced Surgical Techniques*, **14**, 191–195.
6. Fager, P.J. and Per von, W. (2004) The use of haptics in medical applications. *The International Journal of Medical Robotics and Computer Assisted Surgery*, **1**, 36–42.
7. Hu, T., Tholey, G., Desai, J.P., and Castellanos, A.E. (2002) Evaluation of a laparoscopic grasper with force feedback. *Surgical Endoscopy*, **18**, 863–867.
8. Dargahi, J. and Najarian, S. (2004) Human tactile perception as a standard for artificial tactile sensing – a review. *International Journal of Medical Robotics*, **1**, 23–35.
9. van Beers, R.J., Sittig, A.C., and Gon, J.J.D. (1999) Integration of proprioceptive and visual position-information: an experimentally supported model. *Journal of Neurophysiology*, **81**, 1355–1364.
10. Cavusoglu, M.C., Williams, W., Tendick, F., and Sastry, S.S. (2003) Robotics for telesurgery: Second generation Berkeley/UCSF laparoscopic telesurgical workstation and looking towards the future applications. *Industrial Robot, Special Issues on Medical Robotics*, **30**, 22–29.
11. Garcia-Ruiz, A., Smedira, N.G., Loop, F.D. *et al*. (1997) Robotic surgical instruments for dexterity enhancement in thoracoscopic coronary artery bypass graft. *Journal of Laparoendoscopic & Advanced Surgical Techniques Part A*, **7**, 277–283.
12. Dong-Soo, K., Ki Young, W., Se Kyong, S. *et al*. (1998) Microsurgical telerobot system. Proceedings of the 1998 IEEE/RSJ International Conference on Intelligent Robots and Systems, 1998, vol. 2, pp. 945–950.
13. Thompson, J., Ottensmeier, M., and Sheridan, T.B. (1999) Human factors in telesurgery: effects of time delay and asynchrony in video and control feedback with local manipulative assistance. *Telemed Journal*, **5**, 129–137.
14. Okamura, A.M. (2004) Methods for haptic feedback in teleoperated robot-assisted surgery. *Industrial Robot: An International Journal*, **31**, 499–508.
15. Kitagawa, M., Dokko, D., Okamura, A.M., and Yuh, D.D. (2005) Effect of sensory substitution on suture-manipulation forces for robotic surgical systems. *The Journal of Thoracic and Cardiovascular Surgery*, **129**, 151–158.
16. Marohn, C.M.R. and Hanly, C.E.J. (2004) Twenty-first century surgery using twenty-first century technology: Surgical robotics. *Current Surgery*, **61**, 466–473.
17. Maurin, B., Piccin, O., Bayle, B. *et al*. (2004) A new robotic system for CT-guided percutaneous procedures with haptic feedback. Presented at the International Congress Series, 2004.
18. Czibik, G., D'Ancona, G., Donias, H.W., and Karamanoukian, H.L. (2002) Robotic cardiac surgery: present and future applications. *Journal of Cardiothoracic and Vascular Anesthesia*, **16**, 495–501.
19. Awad, H., Wolf, R.K., and Gravlee, G.P. (2002) The future of robotic cardiac surgery. *Journal of Cardiothoracic and Vascular Anesthesia*, **16**, 395–396.
20. Howe, R.D. and Matsuoka, Y. (1999) Robotics for surgery. *Annual Review of Biomedical Engineering*, **1**, 211–240.
21. Kennedy, C., Hu, T., Desai, J. *et al*. (2002) A novel approach to robotic cardiac surgery using haptics and vision. *Cardiovascular Engineering*, **2**, 15–22.

22. Braun, E.U., Mayer, H., and Knoll, A. (2008) The must-have in robotic heart surgery: haptic feedback, in *Medical Robotics* (ed. V. Bozovic), I-Tech Education and Publishing, Croatia, pp. 9–20.
23. Feder, B.J. (2008) Prepping Robots to Perform Surgery. www.nytimes.com/2008/05/04/business/04moll.html, accessed 2012.
24. Payne, T.N. and Dauterive, F.R. (2008) A comparison of total laparoscopic hysterectomy to robotically assisted hysterectomy: surgical outcomes in a community practice. *Journal of Minimally Invasive Gynecology*, **15**, 286–291.
25. Titan Medical INC (2011) Titan Medical's Investor Presentation. www.titanmedicalinc.com, accessed 2012.
26. The Advisory Board Company (2011) A Competitor for Intuitive's da Vinci Robot? www.advisory.com, accessed 2012.
27. Kim, V.B., Chapman, W.H.H. III, Albrecht, R.J. *et al.* (2002) Early experience with telemanipulative robot-assisted laparoscopic cholecystectomy using da Vinci. *Surgical Laparoscopy Endoscopy & Percutaneous Techniques*, **12**, 34–40.
28. Estey, E.P. (2009) Robotic prostatectomy: the new standard of care or a marketing success? *Canadian Urological Association Journal*, **3**, 488–490.
29. Kolata, G. (2010) Results Unproven, Robotic Surgery Wins Converts. www.nytimes.com/2010/02/14/health/14robot.html, accessed 2012.
30. Satava, R.M., Bowersox, J.C., and Mack, M. (2001) Robotic surgery: state of the art and future trends. *Contemporary Surgery*, **57**, 489–499.
31. Cadière, G.B., Himpens, J., Germay, O. *et al.* (2001) Feasibility of robotic laparoscopic surgery: 146 cases. *World Journal of Surgery*, **25**, 1467–1477.
32. Falcone, T., Goldberg, J.M., Margossian, H., and Stevens, L. (2000) Robotic-assisted laparoscopic microsurgical tubal anastomosis: a human pilot study. *Fertility and Sterility*, **73**, 1040–1042.
33. Margossian, H. and Falcone, T. (2001) Robotically assisted laparoscopic hysterectomy and adnexal surgery. *Journal of Laparoendoscopic & Advanced Surgical Techniques*, **11**, 161–165.
34. Marescaux, J., Smith, M.K., Folscher, D. *et al.* (2001) Telerobotic laparoscopic cholecystectomy: initial clinical experience with 25 patients. *Annals of Surgery*, **234**, 1–7.
35. Abbou, C.-C., Hoznek, A., Salomon, L. *et al.* (2001) Laparoscopic radical prostatectomy with a remote controlled robot. *The Journal of Urology*, **165**, 1964–1966.
36. Damiano, R.J. Jr., Tabaie, H.A., Mack, M.J. *et al.* (2001) Initial prospective multicenter clinical trial of robotically-assisted coronary artery bypass grafting. *The Annals of Thoracic Surgery*, **72**, 1263–1269.
37. Mohr, F.W., Falk, V., Diegeler, A. *et al.* (2001) Computer-enhanced 'robotic' cardiac surgery: Experience in 148 patients. *The Journal of Thoracic and Cardiovascular Surgery*, **121**, 842–853.
38. Kappert, U., Cichon, R., Schneider, J. *et al.* (2001) Technique of closed chest coronary artery surgery on the beating heart. *European Journal of Cardio-Thoracic Surgery*, **20**, 765–769.
39. Boehm, D.H., Reichenspurner, H., Gulbins, H. *et al.* (1999) Early experience with robotic technology for coronary artery surgery. *The Annals of Thoracic Surgery*, **68**, 1542–1546.
40. Cisowski, M., Drzewiecki, J., Drzewiecka-Gerber, A. *et al.* (2002) Primary stenting versus MIDCAB: preliminary report-Comparision of two methods of revascularization in single left anterior descending coronary artery stenosis. *The Annals of Thoracic Surgery*, **74**, S1334–S1339.
41. Eckstein, F.S., Bonilla, L.F., Schaff, H. *et al.* (2002) Two generations of the St. Jude Medical ATG coronary connector systems for coronary artery anastomoses in coronary artery bypass grafting. *The Annals of Thoracic Surgery*, **74**, S1363–S1367.
42. Hollands, C.M., Dixey, L.N., and Torma, M.J. (2001) Technical assessment of porcine enteroenterostomy performed with ZEUS™ robotic technology. *Journal of Pediatric Surgery*, **36**, 1231–1233.
43. Hollands, C.M. and Dixey, L.N. (2002) Robotic-assisted esophagoesophagostomy. *Journal of Pediatric Surgery*, **37**, 983–985.
44. Prasad, S.M., Ducko, C.T., Stephenson, E.R. *et al.* (2001) Prospective clinical trial of robotically assisted endoscopic coronary grafting with 1-year follow-up. *Annals of Surgery*, **233**, 725–732.
45. Tholey, G., Chanthasopeephan, T., Hu, T. *et al.* (2003) Measuring grasping and cutting forces for reality-based haptic modeling. Computer assisted radiology and surgery. Presented at the 17th International Congress and Exhibition, London, UK, 2003.
46. Hu, T., Castellanos, A.E., Tholey, G., and Desai, J.P. (2002) Real-Time Haptic feedback Laparoscopic tool for use in Gastro-intestinal Surgery. 5th International Conference on Medical Image Computing and Computer Assisted Intervention (MICCAI), Tokyo, Japan, 2002.

47. Kennedy, C.W., Tie, H., and Desai, J.P. (2002) Combining haptic and visual servoing for cardiothoracic surgery. Proceedings of the ICRA '02. IEEE International Conference on Robotics and Automation, 2002, vol. 2, pp. 2106–2111.
48. Kennedy, C.W. and Desai, J.P. (2003) Force feedback using vision. The 11th International Conference on Advanced Robotics, University of Coimbra, Portugal, 2003.
49. Morimoto, A.K., Foral, R.D., Kuhlman, J.L. *et al*. (1997) Force sensor for laparoscopic Babcock. *Studies in Health Technology and Informatics*, **39**, 354–361.
50. Tozzi, P., Corno, A.F., and von Segesser, L.K. (2002) Sutureless coronary anastomoses: revival of old concepts. *European Journal of Cardio-Thoracic Surgery*, **22**, 565–570.
51. Buijsrogge, M.P., Scheltes, J.S., Heikens, M. *et al*. (2002) Sutureless coronary anastomosis with an anastomotic device and tissue adhesive in off-pump porcine coronary bypass grafting. *The Journal of Thoracic and Cardiovascular Surgery*, **123**, 788–794.

10

Teletaction

Teletaction is the means by which an operator is presented with information about the texture, local shape, and/or compliance of a remotely located object through the use of a tactile display system. The ideal teletaction system is one in which the user is provided with a pattern that is indistinguishable from direct contact.

10.1 Introduction

Although most computer systems are able to provide unrivalled processing capabilities for standard and routine applications, a human is able to control a device in unknown or difficult situations more effectively than any computer program due to possessing greater intuitive feel and superior decision-making capabilities. In order to undertake any computer or robotic processes that involve one or more of the primary senses, data will first be required from devices that replicate the senses of sight, sound, smell, taste, and touch. In order to accomplish this, it is necessary to fabricate a remote sensory device that is able to transmit to the user stimuli that are identical to that which they would experience were they to be in *direct* contact with and/or exposed to the medium or object. For humans, the two most important senses for manipulation tasks are sight and haptics (the sense of touch). The haptic sense itself comprises two modes, namely the kinesthetic sense (force, motion) and the tactile sense (touch) of which the latter includes texture, roughness, softness, temperature, and shape. The main applications for tactile systems are to be found in the teleoperation and virtual environment (VE) milieu. A tactile interface is used to reproduce information such as force (static and dynamic), texture, roughness, temperature, and shape which are important in applications such as telesurgery or when handling fragile objects during a manipulation process. To implement tactile feedback, the problems of tactile transduction, signal processing, tactile display, and human perception must first be considered.

From a biomedical engineering point of view, teletaction is the sensing of a remote biological tissue and then transmission of this cutaneous (tactile-sensed) information to the operator's skin (typically the fingertips) [1, 2]. This practice has already been used in computer-assisted surgery and VEs for training purposes. As previously mentioned,

Tactile Sensing and Displays: Haptic Feedback for Minimally Invasive Surgery and Robotics, First Edition.
Saeed Sokhanvar, Javad Dargahi, Siamak Najarian, and Siamak Arbatani.
© 2013 John Wiley & Sons, Ltd. Published 2013 by John Wiley & Sons, Ltd.

a tactile interface is required to reproduce, as accurately as possible, parameters such as force (static and dynamic), texture, roughness, temperature, and shape. This sense of touch is especially important in surgical applications where the feel of the environment provides knowledge that cannot be obtained by purely visual means [3–5]. In a typical teletaction system, this sensed information is fed into a processing and control unit [6] in which the latter controls the tactile display devices that provide the same sensation as if the skin itself were touching the object.

10.2   Application Fields

Teletaction systems are now being used at an ever-increasing rate within the medical, entertainment, educational, military, telerobotic systems, and consumer electronics sectors [7–9]. However, this paper will confine itself to the use of teletaction in the medical sector, specifically its use in the following contemporary fields: telemedicine, e-health, telepalpation, telemanipulation, and telepresence.

10.2.1   Telemedicine or in Absentia Health Care

Telemedicine generally refers to the use of telecommunication and information technologies in order to provide clinical health care at a distance. It helps eliminate distance barriers and can improve access to medical services that would often not be consistently available in distant rural communities. It is also used routinely to save lives in critical care and emergency situations. Telemedicine is a rapidly developing application of clinical medicine in which medical information is transferred through interactive audiovisual media for the purpose of consulting and, occasionally, for remote medical procedures or examinations. Telemedicine may be as simple as two health professionals discussing a case over the telephone, or as complex as using satellite technology and videoconferencing equipment to conduct a real-time consultation between medical specialists in two different countries. *In absentia* health care is an old practice which, in its very early years, was often conducted using the postal service. Since then, however, there has been a long and successful history of *in absentia* health care which, thanks to modern communication technology, has evolved into what we know today as telemedicine. Telemedicine can be categorized into three main groups, namely: store-and-forward, remote monitoring, and interactive telemedicine.

Store-and-forward telemedicine services collect clinical data, store them, and then forward them to be interpreted later. These systems have the ability to capture and store digital still or moving images of patients, as well as audio and text data. A store-and-forward system eliminates the need for the patient and the clinician to be available at the same time and place. Store-and-forward is therefore an asynchronous, non-interactive form of telemedicine, and is usually employed as a clinical consultation (as opposed to an office or hospital visit).

Remote monitoring enables physicians and other health care providers to be apprised of physiologic measurements, test results, images, and sounds, usually collected in a patient's residence or care facility. Post-acute-care patients, patients with chronic illnesses, and patients with conditions that limit their mobility often require close monitoring and follow-up.

Interactive telemedicine services are real-time clinician–patient interactions that, in the conventional approach, require face-to-face encounters between a patient and a physician or other health care provider. Examples of clinician-interactive services that might be delivered by telemedicine include online office visits, consultations, hospital visits, telephone conversations, and home visits, as well as a variety of specialized examinations and procedures.

10.2.2 Telehealth or e−Health

Electronic health (e-health) is much broader than telemedicine or telehealth. It covers the use of digital data transmitted electronically–for clinical, educational, and administrative applications–both locally and at a distance. Telehealth evolved from telemedicine and, in addition to the curative aspect upon which telemedicine itself is primarily focused, it also concentrates on health promotion and preventive measures, with heavy reliance on technology to provide solutions. Telehealth uses video conferencing and supporting technologies to put patients in touch with health professionals across distances. It is especially useful in remote areas where patients, particularly the elderly and those with young children, would otherwise have to travel long distances to meet health professionals. It also provides access to a wider range of specialist advice and services and delivers faster, more efficient health care by using technology to remove the distance barrier.

In the simplest form of store-and forward-mode, basic vital signs like blood pressure, weight, pulse oximeter, and blood sugar values are monitored and trended for long-term chronic care.

In real-time or interactive mode a telecommunications link allows instantaneous interactive examination by using peripheral devices attached to computers. Examples of real-time services under the aegis of telehealth include audiology, cardiology, dentistry, mental health, neurology, nursing, radiology, and rehabilitation.

In an effort to enhance the real-time telehealth experience, Google Health, a personal health information centralization service (sometimes known as personal health record services), recently began establishing relationships with telehealth providers that allow their users to synchronize the data shared during telehealth consultations with their online health records. In remote monitoring, sensors are used to capture and transmit biometric data. For example, a tele-electroencephalogram monitors the activity of a person's brain and then transmits the data to a specialist in either real-time or store-and-forwardmode. Remote monitoring may include video conferencing, messaging reminder, and surveillance questioning, and/or one or more sensors, such as electrocardiogram, pulse oximetry, vital signs, weight, glucose, and movement or position detectors, all undertaken between sessions with his/her physician. The benefits of telehealth include fewer hospital incarceration periods and emergency room visits as well as a reduced need to attend distant health centers [7].

10.2.3 Telepalpation, Remote Palpation, or Artificial Palpation

Tissue palpation is a conventional method to determine the characteristics and possible anomalies in touched tissues. This technique is not applicable in minimally invasive situations, as the surgeon no longer has direct access to the tissue and must operate solely through instruments. There are reports of surgeons performing laparoscopic surgery who found that the lack of direct access to the tissue by the finger, as well as insufficiency of visual feedback, rendered it impossible for them to feel the presence of tumors in underlying body tissue. This lack of tactile feedback makes certain tasks more difficult, as has been proven by experiments performed by Massimino and Sheridan [8]. Additional experiments demonstrated the importance of tactile feedback such as simple tracking tasks [9], reaction time reduction in target pointing [10], and in degraded visual conditions [8].

One of the surgeon's most important assets is a highly developed sense of touch. Surgeons rely on sensations from the finger tips to guide manipulation and to perceive a wide variety of anatomical structures and pathologies [11]. In current minimally invasive methods, as previously mentioned, the surgeon's perception is limited to visual feedback or force feedback from the handles of their instruments. Various tactile sensors have recently been developed which can be mounted in a probe, or in a surgical instrument, from which any tactile information sensed is transmitted through a controller to a tactile display which reproduces the tactile stimulus. By using these remote palpation devices, the surgeon may get back some of the perceptual and manipulative skills of conventional open-incision surgery. Among the tactile feedback parameters, scientists are also investigating force reflection [12], vibration [13], and small-scale shape recognition [11]. For example, in the case of shape recognition, which is important for many surgical tasks, such as finding tumors, a tactile sensor array in the tip of surgical instrument or probe measures the distribution of stress after contacting the tissue. This data is then processed by a computer using signal-processing algorithms, after which the output data is either shown on a video screen or represented on a tactile display. A schematic remote palpation system is represented in Figure 10.1.

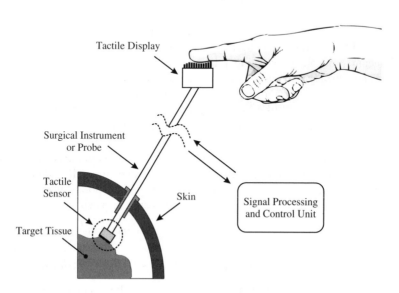

Figure 10.1   Schematic view of remote palpation system

10.2.4   Telemanipulation

Although the human being is very adept at performing manipulative tasks, there are clearly a number of environments in which the safety aspect and location render any such undertakings either impossible or hazardous (such as in nuclear stations, mines, polluted environments, war zones, or in remote environments such as deep sea or outer space). Even the use of automatic robots may not be possible due to safety and reliability issues. Telemanipulation, which is an extension of teleoperation, provides a skilled human operator with the means to perform manipulation tasks using a robot that can, itself, be placed in a situation or environment that would be not possible or not safe for a human [14–16].

Telemanipulation operations, however, cannot be accomplished without visual [17] and haptic telepresence capability, which allows the human teleoperator to experience the same feeling as the robot itself [16, 18–21].

Adding force feedback and tactile sensing in telemanipulation tasks improves task completion time and accuracy. For example, in the Dextrous Telemanipulation project that originated in Stanford University, a telemanipulator system was developed which allows a human to control a dexterous robotic hand in an intuitive manner (Figure 10.2). The master system is centered on the human hand and finger motions. A glove, upon which instruments are attached, is used to measure finger motions and a lightweight exoskeleton force-feedback device provides fingertip-level force feedback to the operator (Figure 10.2a). The slave system consists of a custom-designed two-fingered dexterous robotic hand outfitted with force and tactile sensors (Figure 10.2b). The robot hand is attached to a larger industrial robot for increased workspace.

As an another example, Dextre (also known as Special Purpose Dexterous Manipulator (SPDM) or Canadarm because of Canada's contribution to the International Space Station (ISS)) is the most sophisticated robot ever built as a space handyman with the maintenance mission of keeping the ISS ship-shape (Figure 10.3).

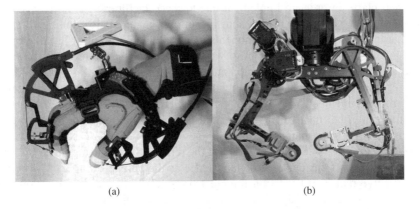

(a)   (b)

Figure 10.2   Dextrous telemanipulation system from Stanford University. (a) The master system consisting of an instrumented glove for finger motion measurement and an exoskeleton for fingertip force feedback [22]. (b) The slave robotic hand with two fingers and fingertip force sensors for relaying environmental interactions (See Plate 18)

Figure 10.3   Dextre, ISS Canada 'hand' (Courtesy of NASA)

10.2.5   Telepresence

Telepresence creates the illusion that remote participants are in the same room to-gether. Telepresence is a term that describes a set of technologies that allows a human to feel presence in a place or have an effect in that place. Prerequisite for having such a feel-ing is that a stimulus of the interaction between the user and the remote place must be pro-vided to the user in order to give the feeling of being in that remote location. As mentioned in the definition of telemanipulation, the user is able to change or influence certain physi-cal aspects within a remote location. In telepresence, all audible and visual aspects within each location are transmitted and received by each location in a multiplex context. Telep-resence is distinct from virtual reality insofar as it refers to a user interacting with another live and real place. In order to implement telepresence facilities, it is necessary to install equipment with suitable audio, video and, if required, manipulation abilities. Of these, it is clear that the audio and video components are preponderant in most installations. From the point of view of manipulation, however, it is evident that a different technology is required, as discussed under telemanipulation. In this section, it was explained that the more closely the robot recreates the manipulation abilities of the human hand, the more the sense of immersion. Robotic devices can vary from being simple one-axis grippers to fully anthropomorphic robot hands. Kinesthetic and tactile telemanipulation refers to a system that provides some sort of haptic feedback to the user who then feels some approximation of the weight, firmness, size, and/or texture of the remote objects manipulated by the robot.

One industry expert has described telepresence as the human experience of being fully present at a live real-world location remote from one's own physical location [23]. As previously mentioned, anybody in attendance at a telepresence gathering is able to prop-agate and receive stimuli from any location. There are many and various examples in which this application is undertaken, including emergency management and security ser-vices, business and industry (B&I), and the entertainment and education industries [23]. In addition to the applications already presented in telemanipulation, telepresence has other applications such as connecting communities, remote inspections, and educational applications, such as providing professional development to teachers for which one of the most effective forms is coaching, or cognitive apprenticeship [24].

One promising new advent in the use of telepresence is its use in performing real-time surgical operations, as demonstrated in Regensburg, Germany in 2002. Also the transfer of haptic (tactile) information has also been demonstrated in telemedicine [25].

10.3   Basic Elements of a Teletaction System

The general elements of a teletaction system are shown in Figure 10.4. As previously stated, a teletaction system is the means by which an operator is presented with information about the texture, local shape, and/or compliance of a remotely located object through the use of a tactile display system. One of the more demanding medical applications of teletaction is the employment and use of robotic laparoscopic telesurgery systems. A tactile sensor is mounted on the end-effector (the laparoscopic instrument) and a tactile display is mounted on the master manipulator (the user interface) which presents information gathered by the tactile sensor to the user. Ideally, the patterns felt by the user would be indistinguishable from direct contact with the environment. The mechanical tactile display needs to generate surface stresses that realistically represent data collected by the tactile sensor. To fully control surface stress, the ideal tactile display system would be an infinite density array of three-DOF actuators.

Generally, teletaction systems are composed of a tactile sensor, a tactile filter, and a tactile display, as shown in Figure 10.4. Most teletaction work has focused its attention on tactile sensors and displays with high spatial but low temporal resolutions [11, 27]. A tactile filter, which is typically a computer with A/D and D/A boards, converts tactile sensor data into tactile display control data. Some concerns regarding tactile filter design, however, are the spatial and temporal sampling differences between the sensor and the display. Also, anomalies exist when attempting to filter strain profile data from the sensor, and convert them to normal and shear force displacement profiles for the display, due to the inability of the filter to completely remove all noise from the system.

One promising use for a teletaction system is in minimally invasive surgery (MIS). Figure 10.5 suggests using a tactile sensor mounted on a catheter to allow a surgeon performing vascular surgery to feel for plaques, branches, or soft spots inside blood vessels. The teletaction system uses a tactile stimulator to create a pattern of stress on a fingertip. This pattern is ideally indistinguishable from direct contact of the finger with the environment.

10.4   Introduction to Human Psychophysics

The term psychophysics suggests a combination of both physics and psychology. More specifically, psychophysics is a science concerning human and animal sensory responses

Figure 10.4   A block diagram representation of a general teletaction system [26]

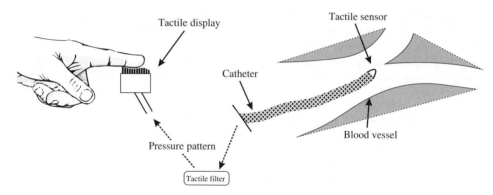

Figure 10.5 An example of teletaction system application in MIS, allowing a vascular surgeon to feel plaques, branches, or soft spots inside blood vessels [1]

to physical and also chemical stimuli. For instance, seeing is a response to light. We may see the light as red or yellow or any other color, or a mixture of colors. It may appear as a short flash or a steady illumination; it may be patterned and appear as stripes, circles, squares, or any other shape. Hearing is a response to any sound, the sense of touch becomes evident from pressure on the skin's surface, smell is derived from odorant substances and taste comes from gustatory chemicals.

To a large extent, psychophysics is a quantitative science. For example, it attempts to describe numerically how fast a sensation grows with the intensity of a physical stimulus. When we increase the intensity of sound, how fast does its loudness increase, by how much do we have to increase the intensity of a light to increase its brightness by a factor of two or three or any other ratio? It also attempts to specify the thresholds in which stimuli can still be detected, such as the smallest pressure on the skin that can be noticed, or the smallest perceptible concentration of an odorant or a gustatory chemical.

Psychophysical processes are involved in almost everything we do. When we compare the height of a building to another by eye, that is psychophysics; when we compare two tones, that is psychophysics also; even when we decide which coffee or tea we prefer, we perform a psychophysical tasting operation. In sports, whenever we throw or catch or hit a ball, subjective psychophysical measurement is involved. We can hardly make a move without running into psychophysics. The term psychophysics was first coined by the nineteenth-century physicist, Gustav Theodor Fechner, who was interested in the relationship between the material and the spiritual. The first psychophysical law was established by Fechner on the basis of the observation of Ernst H. Weber that barely noticeable increments in lifted weights are directly proportional to the base weight [28]. Fechner also established another famous law named after him. With intuitive insight, he decided that subjective impressions grew as logarithms of the physical stimuli that produced them, that is, $S = K \log I$ [29, 30].

Psychophysics ranges from sensory processes, through to sensory memory and short-term memory issues and then to the interaction between sensation and action. The dynamics and timing of human performance are a further important issue within this extended framework of psychophysics. Given the similarity of the various cortical areas in terms of their neuroanatomical structure, it is an important question whether this similarity is paralleled by a similarity of processes. These issues are addressed by state-of-the-art research methods in behavioral research, psychophysiology, and mathematical modeling [31, 32].

Psychophysics, as established by Fechner, aimed at finding pervasive quantitative rules for relationships between the intensity of stimulus and the psychophysical sensation itself. In physics, which derives a tremendous power from them, such rules are called scientific laws. As mentioned, the first psychophysical law was established by Fechner. The observation proved to hold not only for weight, but for other physical variables as well. Although Fechner's logarithmic law was much later proved to be incorrect (notwithstanding its association with Weber's law) it did, nevertheless, demonstrate the importance of Fechner as a physicist and the contribution he made to scientific laws. They point to relationships of great generality from which specific relationships can be derived and, in so doing, they provided the basic structure of this science.

Most recently, psychophysics and physiology together have become essential parts of environmental sciences by telling us how our environments affect us. Psychophysics may be defined nowadays as a science of quantitative relationships between psychological variables and the physical variables that elicit them. 'Physical' is used here as a generic term that includes 'chemical.' Some of the relationships are so intimate that before sufficient instrumentation was developed, the only way people knew about the physical events was through their senses. This is probably the reason why, even today, the same word is used for the physical light as for the sensation of it. The same is true for sound and some other physical variables. We have to be clear in specifying whether we mean the physical variable or its sensation. When we say 'the light is bright' we really mean our sensory impression, not the physical quantity that we can know only through inference. The inference may be quite inaccurate and depend on context variables. Optical illusions are well known.

The laws constitute the backbone of a science from which other relationships can be derived, and the effects of variables producing deviations from the relationships defined by the laws studied. For example, Newton's law applies to objects falling in a vacuum. When an object falls in air due to the effect of gravity, its acceleration is decreased by air resistance. To establish a law empirically, all the variables not specified in the law must be eliminated or their effects determined and accounted for.

In psychophysics, the determination of sensory characteristics occurs indirectly by measuring one stimulus variable as a function of another. For example, the threshold of audibility is determined as a function of sound frequency, the threshold of visibility as a function of the wavelength of light, detectable vibration as a function of vibration duration, and so forth. It is also possible to measure magnitudes of different stimuli producing equal subjective magnitudes such as, for example, sound intensities at two different frequencies that produce equal loudness. While all such measurements have proven to be useful they are, nevertheless, only limited to threshold values, or to subjective magnitudes relative to others specified indirectly in terms of the stimulus values that produce them. Such stimulus-oriented psychophysics provides only an incomplete image of our sensory functioning that most often occurs at stimulus values that only have sufficient strength or quantity to produce a barely perceptible physiological effect (suprathreshold) regardless of specific reference standards.

An essential step in any teletaction system is to first determine the type of stimulus that is to be investigated, such as sampling density or amplitude resolution, so that the appropriate design and system requirements can be formulated from which psychophysical experiments can eventually be conducted. By contrast, psychophysical studies of haptic fidelity apply careful metrics to user task performance during the execution of simple tasks [33]. Psychophysical study of teletaction systems is necessary for several reasons. Firstly, the mechanism of how variations in a tactile display's actuator parameters change the perception of shape and dynamical properties of contact, such as texture and friction, is still not sufficiently understood. Secondly, suitable sensory resolution associated with the teletaction system still needs to be obtained through psychophysical experiments. Thirdly, for the application uniqueness of each teletaction system there is still insufficient data that describes detection thresholds and associated psychophysical functions.

10.4.1   Steven's Power Law

A law relating the objective, instrument-measured intensity of a stimulus to its intensity as perceived by a human, was enunciated in 1957 by Stanley S. Stevens. It addresses the same question that Fechner did almost 100 years earlier, but while Fechner postulated that the perceived intensity is always related logarithmically to the physical intensity, Steven's law says the magnitude of the perceived intensity is related to the magnitude of the physical intensity raised to some power and is sometimes simply called the Power Law. Stevens' power law is often considered to supersede Fechner's law (known as the Weber–Fechner law) upon which he expanded and included a wider range of sensations. In it he was able to demonstrate, more convincingly than his predecessors, that mutually consistent ratio measurements of loudness and brightness were possible. The principal methods used by Stevens to measure the perceived intensity of a stimulus were magnitude estimation (MA) and magnitude production (MP). In magnitude estimation with a standard, the experimenter presents a stimulus called a standard and assigns it a number called the modulus. For subsequent stimuli, subjects report numerically their perceived intensity relative to the standard so as to preserve the ratio between the sensations and the numerical estimates (e.g., a sound perceived twice as loud as the standard should be given a number twice the modulus). In ME without a standard (usually just ME), subjects are free to choose their own standard, assigning any number to the first stimulus, and all subsequent ones with the only requirement being that the ratio between sensations and numbers is preserved. In magnitude production a number and a reference stimulus is given and subjects produce a stimulus that is perceived as that number times the reference. Also used is *cross-modality matching*, which generally involves subjects altering the magnitude of one physical quantity, such as the brightness of a light, so that its perceived intensity is equal to the perceived intensity of another type of quantity, such as warmth or pressure. The initial experiments were performed on hearing, in which an individual was given a reference standard consisting of a tone at a predetermined intensity and a number to express its subjective loudness magnitude. The subject was instructed to assign numbers to subsequently presented tones in proportion to their loudness magnitudes relative to the standard. The numbers proved to follow a power function law for which the general form is:

$$\psi\,(I) = kI^a [34]$$

where I is the magnitude of the physical stimulus, $\psi(I)$ is the psychophysical function relating to the subjective magnitude of the sensation evoked by the stimulus, a, is an exponent that depends on the type of stimulation, and k is a proportionality constant that depends on the type of stimulation units used. Table 10.1 lists exponents reported by Stevens.

Repetition of the experiment on several other observers produced similar responses. An analogous result was obtained when light flashes were substituted for the tonal stimuli. The subjective brightness magnitudes, as expressed by assigned numbers, followed a similar power function. Stevens decided that he may have found a general principle for the relationships between sensory stimulus intensities and the subjective sensation magnitudes they evoked. Because the relationships followed power functions, he designated

Table 10.1 Exponents reported by Stevens [34]

Continuum	Exponent (a)	Stimulus condition
Loudness	0.67	Sound pressure of 3000 Hz tone
Vibration	0.95	Amplitude of 60 Hz on finger
Vibration	0.6	Amplitude of 250 Hz on finger
Brightness	0.33	5° target in dark
Brightness	0.5	Point source
Brightness	0.5	Brief flash
Brightness	1	Point source briefly flashed
Lightness	1.2	Reflectance of gray papers
Visual length	1	Projected line
Visual area	0.7	Projected square
Redness (saturation)	1.7	Red-gray mixture
Taste	1.3	Sucrose
Taste	1.4	Salt
Taste	0.8	Saccharin
Smell	0.6	Heptane
Cold	1	Metal contact on arm
Warmth	1.6	Metal contact on arm
Warmth	1.3	Irradiation of skin, small area
Warmth	0.7	Irradiation of skin, large area
Discomfort, cold	1.7	Whole body irradiation
Discomfort, warm	0.7	Whole body irradiation
Thermal pain	1	Radiant heat on skin
Tactual roughness	1.5	Rubbing emery cloths
Tactual hardness	0.8	Squeezing rubber
Finger span	1.3	Thickness of blocks
Pressure on palm	1.1	Static force on skin
Muscle force	1.7	Static contractions
Heaviness	1.45	Lifted weights
Viscosity	0.42	Stirring silicone fluids
Electric shock	3.5	Current through fingers
Vocal effort	1.1	Vocal sound pressure
Angular acceleration	1.4	5 s rotation
Duration	1.1	White noise stimuli

the principle as the Power Law, which has since been confirmed by many experimenters in many experiments performed in many sense modalities. Next to the Weber Law that withstood the test of time for almost two centuries, it is the best-documented general relationship of psychophysics. Because it may be considered as the answer to Fechner's fundamental question of the relationship between the 'spiritual' and the 'material,' to use Fechner's language, Stevens regarded the Power Law as the Psychophysical Law. In his monograph, the view is accepted that the Power Law is the most fundamental law of psychophysics, and he designates it as the First Law of Psychophysics. Nevertheless, additional laws are possible and may have considerable usefulness [34].

10.4.2   Law of Asymptotic Linearity

Before the Power Law was firmly established, methods of measuring psychological quantities had to be developed. The original method introduced by Stevens to measure loudness and brightness, which he called 'magnitude estimation,' proved partially misguided and produced the correct results only by fortuitous happenstance. Stevens and his coworkers soon discovered that the exact functions relating to the psychological magnitudes of the underlying stimulus magnitudes depended upon the designated reference standards. As it transpired, the 'best' power functions were obtained when the observers were allowed to choose the standards themselves. This discovery suggested that the observers did not strictly obey the rules of ratio scaling, which allows for entirely arbitrary reference standards, but, to some degree, attached absolute values to numbers. A more systematic investigation of the effects of reference standards was performed by J. Zwislocki and Rhona P. Hellman [35, 36]. The investigation led to the conclusion that experimental observers do not use numbers in a relative way, depending on chosen units, but rather in an absolute way. In other words, probably because of the way they are used in everyday life and the way children use them when learning, they acquire absolute subjective values. For example, children learn numbers by counting objects such as pebbles or pencils, so coupling numbers to perceived objects occurs early in life and complies with numeric rules in which numbers have absolute recurring values. When asked to assign numbers to subjective impressions of line length or to loudness, adults and children produced the same absolute functions within the range of the numerals they knew. If numbers acquire absolute subjective values, ME becomes a matching operation. The subjective values of numbers are matched to the subjective values of whatever variable is being scaled. Because Stevens' method of ME appeared to produce biased results due to asymmetry, he introduced a complementary method in which numbers were given by the experimenter, and the observers had to find matching sensation magnitudes that they produced by manipulating appropriate instrumental controls. He called this method 'MP'. In the methods of scaling subjective magnitudes developed by Zwislocki and Hellman, the numbers are assumed to have absolute subjective values. Consequently, they call what started as ME, 'absolute magnitude estimation' (AME), and what started as MP, 'absolute magnitude production' (AMP). The designations conserve Stevens' tradition but are not completely accurate because both are regarded as matching operations. The methods have opened a wide world of subjective magnitudes to measurement in spite of objections by staunch opponents who do not believe that they constitute legitimate measurements. However, the mutually consistent results tend to cast doubt on these objections. Sensation magnitudes almost generally follow power functions of adequate stimulus variables, except at very low values of these variables. When the threshold approaches within which the variables can be detected, their subjective magnitudes converge on either direct proportionality to the stimulus intensity or some other related variable. Line length squared would be an example of such a variable. For sufficient lengths, the subjective line length tends to be directly proportional to the physical line length. As surprising as it may appear, this is no longer true for very short thin lines that become somewhat difficult to see. According to measurements by Sanpetrino (unpublished result), the subjective line length then becomes proportional to the square of the physical line length. This phenomenon can be explained by the physiological noise that is added to the visual line image. In agreement with the theory of signal detectability, such a process can be expected to take place near the threshold in which all sensory stimuli can be detected and would be consistent with a linear relationship of subjective magnitudes to stimulus intensity.

As previously mentioned, according to this law, all subjective magnitudes grow linearly with the intensities of the stimuli that evoke them near their detectable thresholds. The relationship was first discovered during auditory measurements of Stevens' Power Law and is described here initially for loudness, then generalized. When sufficiently small stimulus magnitudes were included, the resulting loudness curves deviated from the Power Law and, on double-logarithmic coordinates when bent downward, became gradually steeper. A typical example is shown in Figure 10.6, where loudness magnitudes of a 1000 Hz tone are plotted over sound-intensity abscissas expressed in decibels [37]. The solid curve has been determined by the method of ME based on two reference standards, as described in Chapter 9 [36]. The slanted crosses show averages of 12 studies computed by Robinson [38], in which various methods were used. Filled circles indicate the data of Stevens [39] obtained by ME with the reference standards chosen by the observers themselves; open symbols and filled triangles, the data of Scharf and Stevens [40] obtained by ME with a designated reference standard and by halving and doubling; the vertical crosses, the data determined by Feldtkeller *et al.* [41] with the help of the same method. The excellent agreement between the various sets of data and the curve suggests that the curve accurately represents the loudness of a 1000 Hz tone as a function of its intensity. Of particular interest is the asymptotic convergence of the curve on a linear relationship between loudness and sound intensity near the threshold of audibility, as indicated by the straight line having the coordinates of 0.01 at 0 SL (threshold of audibility) and 1 at 20 dB [32].

10.4.3 Law of Additivity

This law states that in ME and MP, subjects that are tested experimentally tend to pair numbers with sensation magnitudes on absolute rather than ratio scales. This implies that not only sensations, but also numbers acquire absolute psychological magnitudes. The specific experiments are performed on loudness and line lengths. The latter reveals that the subjective magnitudes of numbers are formed before the age of six and do not change after that age. It is suggested that the absolute coupling of numbers with sensation magnitudes originates from the concept of numerics where numbers have absolute meanings. Additivity of subjective magnitudes is introduced as the third law by Zwislocki [32].

Nevertheless, the existence of additivity has been demonstrated with scientific certainty in hearing, touch, and vision, although the situation in chemical senses had to be left unresolved [32].

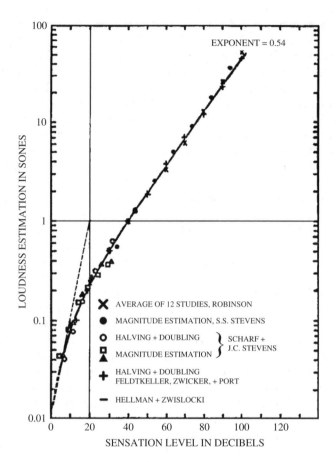

Figure 10.6 A typical binaural loudness function (solid line) determined with two reference standards and compared to the results of five other studies performed with different methods. The intermittent straight line shows a linear relationship between loudness and sound intensity (Modified from Hellman and Zwislocki (1963), © American Institute of Physics [32])

10.4.4 General Law of Differential Sensitivity

This fourth and last law concerns the extent to which increments in audible intensity can be detected. One of the oldest laws in psychophysics is Weber's law, which states that intensity increments that are barely detectable are, in magnitude, directly proportional to the base intensity. This law is often expressed as the Weber fraction consisting of the ratio between the just-noticeable increment and the base magnitude which tends to have a constant value, except at very low stimulus values, where it rapidly increases. In hearing, for pure tones, and in vibrotaction (the response of tactile nerve endings to varying forces on the skin and to oscillatory motion of the skin), for any stimuli, the value tends to decrease slowly as the base intensity increases and approximates to Weber's law.

In more recent times, paradoxical properties of Weber's law have been discovered within the auditory fraternity. When measured by means of just detectable intensity increments or the difference between two intensity increments, Weber's fraction has been shown not to depend on the rate of growth of loudness with stimulus intensity but only on the loudness itself. When measured as the standard deviation of the variability of loudness, it was found to depend on the slopes of the functions in a predictable but complicated way. Counter-intuitively, it depended not only on the slope of the loudness function of the ear, in which the sound intensity was varied, but also on the slope of the loudness function of the contralateral ear. Zwislocki found it possible to describe the differential intensity sensitivity in all its methodological variations by one simple mathematical equation, which he suggests as being an expression of a general law of differential intensity sensitivity [32].

10.5   Psychophysics for Teletaction

Work is still required in order to bring human tactile perception in teletaction systems to the same level as currently exists in auditory and vision systems, although parameters such as sensor density, spatial, and temporal frequency response and tactile sensing sensitivity have been studied in some depth [42–45]. Teletaction displays at an approximate density of 70 sensors per cm², and which map information that stimulate the slowly adapting Type I (SA I) mechanoreceptors, have been discussed [26, 46]. Neurophysiological studies suggest that the SA I mechanoreceptors are most important in small-scale perception [47], have a receptive field diameter of 3–4 mm, and a frequency range of 2–32 Hz [48]. Spatial resolution tests performed by static stimulation applied directly to the skin show that the 75% thresholds for gap detection and grating detection are 0.87 and 1.0 mm, respectively [49]. Other studies show enhanced detection of surface roughness by reducing shear stress information [50]. Orientation detection increased significantly when subjects used a piece of paper between the finger and a 0.0127 mm × 3 mm ridge covered by a 0.5 mm smooth card [51]. The dynamic response of the human finger to objects with and without surface roughness was analyzed with finite element modeling [52]. Performance of shape recognition through a tactile display with different pin spacing has also been studied [45]. Amplitude resolution of the human tactile system was measured through several psychophysical experiments by modeling a teletaction system based on a predicted subsurface strain [26]. They studied the effects of shear stress on grating orientation discrimination, and the effects of viscoelasticity on tactile perception for static touch, which resulted in determining the parameters for a teletaction system design (10% amplitude resolution is sufficient for a teletaction system with a 2 mm elastic layer and 2 mm display-stimuli units spacing).

10.5.1   Haptic Object Recognition

Object recognition is the ability to perceive an object's physical properties (such as shape, color, and texture) and apply semantic attributes to the object, which includes the understanding of its use, previous experience with the object and how it relates to others [53].

Identifying everyday common objects is a specific version of the more general perceptual task called pattern recognition which, when recognized, is then assigned to some category of past experiences. For example, a visual display may be called a 'chair.' In general, theories of pattern recognition tend to include a number of assumptions, such as a process of analysis in which sensory systems break down incoming stimulation into component features. From this, and through a process of synthesis, a higher order of system features are built into integrated units. The assumption is that units at one or more of these levels access memory for known categories of patterns, and the best match determines the category to which the incoming stimulus will be assigned. Typically, pattern recognition theories assume that not only are these processes driven by sensory stimulation in a bottom-up fashion, but, also, that they rely on acquired knowledge and that current expectations will be derived in a top-down driven manner [54].

These theoretical assumptions are so sufficiently broad that they can be applied to any sensory modality, for example, vision, touch, audition, or olfaction. But there is no doubt that by far and away the most predominant modality examined in the studies of pattern recognition is vision, followed by speech recognition. In distinct contrast, little work has been devoted to the study of how familiar patterns are recognized by touch [54].

Intuitively, one may feel that the emphasis on vision is appropriate because it is the way objects are typically recognized in the real world. One might believe that there is little role for object recognition by touch in everyday life. It could be argued, however, that the importance of touch in everyday object recognition has not been highlighted to the extent that it deserves and that, in fact, objects are more frequently recognized by touch than people would believe. People often interact with an object without looking at it but, in order to do this and if they wish to acquire any form of recognition, they must first know what the object is. When we dress, we do not have to fixate on buttons to fasten them; the button and buttonhole can be found by touch. When driving a car, we shift gears and adjust knobs without looking at them rather than taking our eyes off the road. Sometimes we are able to locate to certain items by feel alone, such as our wallets or keys, which exemplifies the fact that recognizing common objects by touch is not only something that we *can* do, but something that we *often* do on a reflexive basis [54].

Haptics is a multi-disciplinary field which plays a key role in any scientific multi-disciplinary field that involves touch, neuroscience, virtual reality, and robotics. Before continuing with our discussion on haptic object recognition, let us first briefly consider how visual object recognition is performed. One model of object recognition, based on neuropsychological evidence, provides information that allows us to divide the process into four different stages [55–57].

1. Processing of basic object components, such as color, depth, and form.
2. Grouping similarly shaped components together in order to show their outlines and segregate each visual form on the basis of figure-ground perception.
3. Storing a visual representation in data memory and then matching it against structural descriptions.
4. Obtaining semantic attributes that are applied to the visual representation in order to provide meaning and recognition.

Within each of these stages, there are more specific processes that take place to complete the different processing components.

Visual recognition processing has been typically viewed as a bottom-up hierarchy in which information is processed sequentially with increasing complexity. At the bottom of the processing hierarchy are the lower cortical processors that deal with aspects such as the primary visual cortex; at the top are the higher-level critical processors such as the inferotemporal cortex (IT), where recognition is facilitated [58]. One of the more recognized bottom-up hierarchical theories is that formulated by David Marr in his theory of vision [59]. By way of contrast, an increasingly popular recognition processing theory is that of top-down processing. One model, proposed by Moshe Bar [58], describes a 'shortcut' method in which early visual inputs are sent, partially analyzed, from the early visual cortex to the prefrontal cortex (PFC). Recognition memory can be supported by both the assessment of the familiarity of an item and by recollection of the context in which an item was encountered. Some have hypothesized that the PFC disproportionately contributes to recollection, whereas an alternative view is that the PFC contributes to both recollection and familiarity. Possible interpretations of the crude visual input is generated in the PFC and then sent to the IT, subsequently activating relevant object representations which are then incorporated into the slower, bottom-up process. This 'shortcut' is meant to minimize the amount of object representations required for matching, thereby facilitating object recognition [58]. Lesion studies have supported this proposal with findings of slower response times for individuals with PFC lesions, suggesting use of only the bottom-up processing [60].

A significant aspect of object recognition is that of object constancy, that is to say the ability to recognize an object across varying viewing conditions. These varying conditions include object orientation, lighting, and object variability (size, color, and other within-category differences). For the visual system to achieve object constancy, it must be able to extract a commonality from the object descriptions from different viewpoints and the retinal description [61]. Several theories have been generated in order to provide insight as to how object constancy may be achieved for the purpose of object recognition, including *viewpoint-invariant, viewpoint-dependent*, and *multiple views* theories.

Viewpoint-invariant theories suggest that object recognition is based on structural information, such as individual parts, allowing for recognition to take place regardless of the object's viewpoint. Accordingly, recognition is possible from any viewpoint, as individual parts of an object can be rotated to fit any particular view [62]. This form of analytical recognition requires little memory as only structural parts need to be encoded which, in turn, can produce multiple object representations through the inter-relations of these parts and mental rotation [62]. Therefore, storage of multiple object viewpoints is not required in memory.

Viewpoint-dependent theories suggest that object recognition is affected by the viewpoint at which it is seen, implying that objects seen from novel viewpoints reduce the accuracy and speed of object identification [63]. This theory of recognition is based on a more holistic system rather than by parts, suggesting that objects are stored in memory with multiple viewpoints and angles. This form of recognition requires a lot of memory as each viewpoint must be stored. Accuracy of recognition also depends on how familiar the object is from the observed viewpoint [62].

The *multipleview* theory proposes that geometric constraints govern the projection of points, lines, curves, and surfaces in multiple images. Object recognition lies on a viewpoint continuum where each viewpoint is recruited for a different type of recognition. At one extreme of this continuum, viewpoint-dependent mechanisms are used for within-category discriminations whereas, at the other extreme, viewpoint-invariant mechanisms are used for the categorization of objects [63].

Some models for object recognition have been suggested such as the 3D model representation, proposed by Marr and Nishihara [64], who state that object recognition is achieved by matching 3D model representations obtained from the visual object with 3D model representations stored in memory. The 3D model representations obtained from the object are formed by first identifying the concavities of the object, which separate the stimulus into individual parts, and then determining the axis of each individual part of the object. Identifying the principal axis of the object assists in the normalization process, via the mental rotation that is required, because only the canonical description of the object is stored in memory. Recognition is acquired when the observed object viewpoint is mentally rotated to match the stored canonical description [64].

An extension of Marr and Nishihara's model, the recognition by components theory suggested by Biederman [65], proposes that the visual information gained from an object is divided into simple geometric components, such as blocks and cylinders, also known as 'geons' (geometric ions), which are then matched with the most similar object representation that is stored in memory to provide the object's identification (see Figure 10.7) [65]. A small number of geons, appropriately arranged in space and relatively sized, is proposed to be sufficient to compose any familiar object. The process of pattern recognition consists, in part, of extracting edges, determining geons from their spatial layout and combining them into an object, then comparing that object to representations in memory corresponding to object categories.

The basic ideas about edge extraction and combining the elements of object images from edges are widely accepted in theories of visual pattern identification. Applying such a model to haptic pattern recognition is problematical, however, because the haptic system is simply not very good at extracting information about the spatial layout of edges. In a haptic display, three types of edges might be presented: (i) 2D patterns in the range of the fingertip scale, (ii) 2D patterns extending beyond the fingertip scale, and (iii) contours of a fully 3D object. Examples of 2D representations are embossed numerals or alphabetic characters and raised line drawings used as illustrations for the blind. The ability to encode edges from patterns lying under the fingertip was systematically investigated by Loomis [66], who evaluated the ability of both sighted and blindfolded people to identify letters and digits that protruded 6 mm from the surface of the larger dimension.

Besides the kinesthetic and cutaneous groupings, haptics can be categorized as being both active and passive, as shown by Gibson [67, 68]. Although other dividing lines exist, it is these that are the two most important in which passive sensing concerns the analysis

Figure 10.7 How objects can be broken into geons

of static tactile data, whereas active sensing is used when motion is involved. Okamura and Cutkosky [69] stated that motion is required when features, particularly those which are small, cannot be sensed accurately through static touch. One example of this is the evident fact that sliding fingers against any edge will yield far more information about an object's sharpness than can ever be determined by static contact alone. A dexterous hand can manipulate an object and retrieve information on its properties which would otherwise be impossible to determine. For example, by tilting an object, its weight, or center of gravity, can be estimated and by running a finger over its surface, its friction and texture can be approximated, and so on.

From Gibson's point of view and his experience with patterns and objects, haptics does not only consist of kinesthetic and cutaneous sensing receptor surfaces; they are actors and perceivers in the real world. Active touch differs from passive touch in the intentionality of our exploratory behaviors. The distinction is somewhat different from the older reafference/exafference separation in which the former is the self-stimulation of an animate life form as a result of the movements of its own body whereas the latter is due to the result of external factors. Gibson thought that active touch revealed things in either a real, or at least imaginary, way whereas passive touch did not commit itself to any immediate or tangible feeling.

Klatzky and Lederman [70] categorized the properties of the object material in terms of texture, compliance, apparent temperature (due to heat flow), and geometric proportions (size and shape). Indeed, in order for any haptic object identification system to be deemed worthwhile and useful, it must first demonstrate that it is able to provide all this information. Lederman and Klatzky conducted studies that directly addressed the availability of material properties under haptic exploration of objects [71]. The procedure was based on a paradigm from vision, called visual search, as adapted by Treisman and Gormican [72].

In a visual display, it is relatively straightforward to vary the number of items by adding distracters to, or subtracting them from, the field of view. It is however, less obvious as to how the number of items in a haptic display can be varied and it makes the investigation of haptic identification more difficult.

In work conducted by Klatzky and Lederman [73], object properties are divided into four sets: (i) material discrimination (rough surface vs. smooth surface or warm vs. cool surface); (ii) discrimination of a flat surface from a surface with an abrupt spatial discontinuity, such as a raised bar; (iii) discrimination of two- or three-dimensional spatial layout, such as whether a raised dot was on the left or right of an indentation, and (iv) discrimination between continuous three-dimensional contours, such as a curved surface as opposed to a flat surface.

Together, these various research findings suggest that the role of material in haptic object identification could contribute substantially to the high level of performance that is observed. However, in order for material information about the stimulus object to be important in identification, the representation of objects in memory must also incorporate material information that can be matched with the stimulus.

Material properties could be used to represent and identify an object, although one problem with this idea is that the name given to the object is primarily dependant on its shape and geometric property. For example, it is well known that people use naming to divide objects into categories whose members share attributes, and that shape is a particularly important attribute when an object is categorized by its most common name [54].

10.5.2   Identification of Spatial Properties

Shape–can be studied as a whole complex body or by decomposing its constitutive elements (edges, angles, curves, etc.). Although the haptic discrimination of shape is generally very much less efficient than visual discrimination (considering speed and number of errors), the processing modes of the haptic and visual modalities appear similar in numerous studies. Thus, both modalities are sensitive to the same dimensions of differentiation [74] and to the same effect of complexity, evaluated by the number of sides of each shape [75, 76]. This similarity has been confirmed by Garbin and Bernstein [77] and Garbin [78] who, through studies of multi-dimensional scaling, found that the same shape attributes (size, symmetry, and complexity) determine the subjective similarity of shapes in vision and haptics. But only 67% of the stimuli occupy the same position on the visual and haptic scales, in adults and at the age of six to seven years.

However, in other works, differences do appear between the visual and the haptic processing of shape. Thus, due to the sequential nature of manual exploration and the possibility of modifying the size of the tactual perceptual field at will, touch is less sensitive than vision to the Gestalt laws of organization and spatial configuration. For more information, refer to Chapter 5 which deals with a study of the sensitivity of touch to the law of proximity [79]. On the other hand, Lakatos and Marks [80] found that when adults were asked to make either visual or haptic similarity judgments on pairs of geometrical forms differing in local features or global shape, the pairs with comparable global shape, albeit with different local features, were judged to be less similar by touch than by vision. This differential effect tended to decrease over time and was not related to the haptic exploratory procedures used by the subjects.

In the same vein, the facilitating effect of symmetry, particularly vertical symmetry which is always present in vision (e.g., [81–83]), is not observed in haptic shape perception. Walk [83], and more recently Ballasteros, Manga, and Reales [84], found that 2D non-symmetrical shapes were detected more quickly using haptic methods and with fewer errors. Conversely, with 3D objects, an advantage appeared for symmetrical shapes in touch as well as in vision.

To interpret these results, Ballasteros, Millar, and Reales [85] assumed that the reference frames available in the tasks presented induced a specific spatial organization of the stimulus and were responsible for the facilitating effect of symmetry. Thus, several reference frames are available when manipulating 3D objects with both hands, such as the body's median (Z) axis, the gravitational vertical, and the reference to the position of the hand itself in relation to the head or trunk. But there are far fewer exterior perceptive cues available for small, drawn figures, which are less well discriminated than 3D objects without training. Sensory data in this case can only be related to a body-centered reference frame [86]. As a result, reference to an egocentric Z axis should be easier when the two index fingers are placed on either side of this axis in a two-handed exploration than in a one-handed exploration in which the active index finger is not previously positioned in the body's median axis. If such is the case, the facilitating effect of symmetry in the two-handed condition and the absence of this effect in the one-handed condition should be observed. Ballasteros, Millar, and Reales [85] obtained this result in a task in which the subject was asked to judge if small, drawn figures (2 × 2 cm), symmetrical or not, according to the vertical, horizontal, or oblique axes, are open or closed. Fewer errors and shorter response times were observed when the two index fingers explored rather than when only one was used. According to the authors, this indirect measure presents the advantage of indicating that, as in vision, the effect of symmetry is primitive and incidental in so far that it appears in the initial stages of perceptual encoding and without voluntary research.

Thus, the facilitating effect of symmetry depends on the availability of spatial reference systems. It should be noted, however, that Ballasteros *et al.* [85] obtained facilitation in the two-handed condition only with open figures and not with closed figures, which gave rise to unclear results.

Orientation – the spatial orientation of a stimulus is always relative to a reference axis, so the vertical and horizontal axes always correspond to the direction of gravity and the visual horizon, respectively, and so form a reference frame in 2D space, where all other orientations are said to be oblique.

Regardless of the age of a person, their perception of orientation is more accurate in vision than it is in haptics, and this accuracy improves with time. In haptic orientation perception, systematic anisotropy (a better perception of vertical and horizontal than oblique orientations) in vision is not always observed. This 'oblique effect' [87] is present in all ages for most visual tasks. It becomes more evident when the difference is made between the performances observed for vertical and horizontal orientations (which most often do not differ and are grouped together) and those for oblique orientations (also grouped together when they do not differ). Because it is a question of difference, this oblique effect may appear whatever the overall accuracy of responses. For example, in a reproduction task, the same subjects make average errors in vision and tactile sensation of 1 and 3.8° respectively for the vertical orientation and of 3.6 and 5.7° for oblique orientations [88]. In vision, the processes responsible for the systematic presence of the oblique effect are influenced by multiple factors which operate on different anatomo-functional levels of the visual system according to the nature of the task [89, 90]. In the haptic modality, as will be seen, the existence of an oblique effect *per se* is debated.

Line Parallelism – Kappers [91, 92] and Kappers and Koenderink [93] investigated the haptic perception of spatial relations in both horizontal and mid-sagittal planes. In a simple task, where blindfolded participants were asked to rotate a test bar in such a way that it felt parallel (in physical space) to a reference bar, huge systematic subject-dependent deviations were found. Observers systematically produced deviations of up to 90°, thus setting the two bars perpendicular! This effect was very robust and remained so even after visual inspection and various methods of feedback. Unimanual and bimanual, and also pointing and co-linearity measurements yielded similar results. These results confirmed that haptic space is not Euclidean, contrary to common expectation. The current hypothesis of the authors of these studies is that the deviations reflect the use of a reference frame that is a weighted average of egocentric and allocentric reference frames. Cuijpers, Kappers, and Koenderink [94, 95] performed similar parallelity and collinearity experiments in vision. That visual space is not Euclidean was already established early in the previous century, but these new studies provided extra evidence. For example, bars at eye height separated by a visual angle of 60° have (depending on the participant) to differ by 20° in orientation in order to be perceived as being parallel, which defies conventional explanations.

Length – perception is more accurate in vision than in haptics, especially in the youngest subjects [96]. In the haptic modality, the results observed in the literature are very variable. To understand them, the different methodologies used in the various studies will be examined.

Localization – perception is more accurate in vision than in haptics [97]. As for the study of length perception, the different approaches used for studying this type of haptic perception will be presented.

10.5.3   Perception of Texture

In a wider sense, all physical properties defining a surface's micro-structure are included in the term 'texture,' including roughness, hardness, and elasticity [71] although only roughness, and to a lesser extent hardness, have been studied.

Tactile texture perception is as efficient as visual perception, and sometimes even surpasses it for the extremely fine textures of abrasive papers (ranging from 1000 to 6000 grit, according to standard norms).

10.5.4   Control of Haptic Interfaces

In contact tasks involving finite impedances, either displacement or force can be viewed as being the control variable, with the other being a display variable, depending on the control algorithms employed. However, consistency among free-band motions and contact tasks is best achieved by viewing the position and motion of the band as the control variable, and the resulting net force vector and its distribution within the contact regions as the display variables.

The concept of teleoperation has evolved to accommodate not only manipulation at a distance, but manipulation across barriers of scale and in VEs, with applications in many areas. Furthermore, the design of high-performance force-feedback teleoperation masters has been a significant driving force in the development of novel electromechanical or 'haptic' computer-user interfaces that provide kinesthetic and tactile feedback to the computer user. Since haptic interfaces/teleoperator masters must interact with an operator and a real or virtual dynamic slave that exhibits significant dynamic uncertainty, including sometimes large and unknown delays, the control of such devices possess significant challenges.

The teleoperation controller should be designed with the goal of ensuring stability for an appropriate class of operator, and environment models, as well as satisfying an appropriately defined measure of performance, usually termed as transparency.

A commonly used teleoperation system model, with five interacting subsystems, is shown in Figure 10.8 [98, 99]. The master manipulator, controller, and slave manipulator

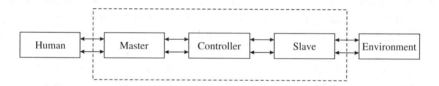

Figure 10.8   Teleoperation system model

can be grouped into a single block representing the teleoperator, as shown by the dashed line. For n-DOF manipulation, the teleoperator can be viewed as a $2n$-port master–controller–slave (MCS) network terminated at one side by an n-port operator block and at the other by an n-port environment network. The force–voltage analogy is more often used than its dual to describe such systems [98, 99]. It assigns equivalent voltages to forces and currents to velocities. With this analogy, masses, dampers, and stiffness correspond to inductances, resistances, and capacitances, respectively.

For passive operator and environment blocks, a sufficient condition for stability is passivity of the teleoperator (e.g., [98]). For the operator to be able to control the slave, a kinematic correspondence law must be defined. In position control mode, this means that the unconstrained motion of the slave must follow that of the master module in some predefined or programmable scaling.

Teleoperation system transparency can be quantified in terms of the match between the mechanical impedance of the environment encountered by the slave and the mechanical impedance transmitted to, or felt by, the operator at the master [99, 100], with the requirement that the position/force responses of the teleoperator master and slave be identical [101].

In spite of the significant amount of research in the area of teleoperation, there are still very few applications in which the benefits of transparent bilateral teleoperation have been clearly demonstrated, even though some areas have great potential, including teleoperated endoscopic surgery, micro-surgery, or the remote control of construction, mining, or forestry equipment. Whether this is due to fundamental physical limitations of particular teleoperator systems or to poorly performing controllers is still not clear. From this perspective, probably the single most important challenge ahead is a better understanding of the limits of performance of teleoperation systems. Toward this goal, it would be useful to have a benchmark experimental system and tasking criteria to be completed, for which various controllers could be tested. Unfortunately, it would be very difficult to do this entirely through simulation due to the fact that the dynamic algorithms necessary to develop a reasonable array of tasks would be just as much under test as the teleoperation control schemes themselves. Furthermore, the minimum number of degrees of freedom for reasonably representative tasks would have to be at least three, for example, planar master/slave systems.

Specific improvements could be made to the fixed teleoperation controllers designed via conventional loop shaping or parametric optimization. In particular, a class of operator impedances that is broader than single fixed impedance, but narrower than all passive impedances, should be developed with associated robust stability conditions. Since the control design problem was formulated as a constrained 'semi-infinite' optimization problem, different algorithms could be tested or new ones developed. Like many other multi-objective optimal control problems, robust teleoperator controller design problems are likely to be hard to solve.

There seems to be much promise in the design of adaptive bilateral teleoperation controllers with relatively simple and physically motivated structures. In particular, indirect adaptive schemes based on Hannaford's architecture [99] are likely to succeed. Whereas fast or nonlinear environment identification techniques are necessary in order to accommodate contact tasks, they are, however, difficult to develop, whereas operator dynamic identification seems to be quite feasible [102]. Some of the difficulties encountered in developing identification algorithms may be circumvented by the use of dual hybrid teleoperation, or newly developed variants that are not based on orthogonal decomposition of the task space into position- and force-controlled spaces. Another interesting research area is the automatic selection of the position and force-controlled subspaces.

10.6 Basic Issues and Limitations of Teletaction Systems

Teletaction has been defined as being the transmission of cutaneous information from a remote tactile sensor to an operator's skin through a tactile filter and display. The main goal of teletaction is to obtain a realistic sense of touch about the shape, hardness, or texture of an object as if touching it directly. As discussed before, teletaction is a branch of haptic feedback with another branch consisting of force or kinesthetic feedback. Most force feedback apparatuses do not provide information about texture, shape, and local compliance. On the other hand, this information is very important in applications such as surgery, where the feel of the environment provides knowledge that cannot be obtained by purely visual means. But, for such critical applications, a teletaction system needs to be supplied with exacting prerequisites which can be only provided by studying each element of a system in order to establish that it meets the exacting limits and boundaries required.

Most of research work has, thus far, been dedicated to tactile sensors. Higher resolution, low-power, and robust tactile sensors capable of detecting the environment through touch are of profound interest in robotic and prosthetic applications. The latest advancement in this field, called artificial skin, which is made of nanowires, was developed and introduced by engineers at the University of California [103]. It is the first such sensor made out of inorganic single crystalline semiconductors, which are integrated at large scale into an active matrix backplane on a thin plastic support substrate. This kind of sensor which works as an artificial skin, would help to overcome a key challenge in robotics by adapting the amount of force needed to hold and manipulate a wide range of objects [103].

Due to highly demanding mechanical requirements, progress in teletaction display has been slow. An ideal stimulator requires 50 N cm^{-2} peak pressure, 4 mm stroke and 25 Hz bandwidth, power density of 5 W cm^{-2} with 1 mm^{-2} actuator density [104]. Two types of teletaction systems are required in order to generate identical stress and strain patterns on the fingertips which can be defined, respectively, as being 'stress matching' and 'strain matching.' An alternative method for teletaction is to produce the object's contour so that it contacts the human hand [18]. This method has some limitations, since the surface defects on the tactile sensor are not the same as the object shape [105]. Additionally the shape display makes it difficult to present shear stresses or tensile forces, which may be possible with the strain matching approach. It is hard to build a stiff display which feels like a rigid object, as the elastic layer is still necessary for anti-aliasing purposes that occur in both the sensor and display parts of a teletaction system.

The physical distance between sensor and display, which causes delays in data transfer, is the other main issue of teletaction systems. Time delay is an inherent problem within most network communication and poses serious problems in haptic interaction when a multiplicity of networks is involved. Sheridan and Ferrell [106] started investigating this problem as early as 1963, when they found that time delay results in a loss of the sense of causality between the operators' hands in team operations and action/reaction mismatch in highly dynamic operations. In the presence of transmission delays, force feedback has a strongly destabilizing effect.

Nowadays, high-speed data transmission links, such as Internet 2, make the teletaction process possible over long-distance telemanipulation processes, even for those with more than one manipulator [107].

10.7 Applications of Teletaction

Artificial stimulation may be used for creation and manipulation of virtual objects, and for the enhancement of the remote control of machines and devices. It has been described as 'doing for the sense of touch what computer graphics does for vision' [108].

Some of the many domains to which teletaction can be applied are as follows:

1. minimally invasive surgery (MIS) and minimally invasive robotic surgery (MIRS)
2. teleoperation and telepresence;
3. laboratory prototypes to study the different tactile parameters;
4. sensory substitution;
5. 3D surface generation;
6. Braille systems;
7. entertainment industries.

Since MIS and MIRS is the focus of this book, these applications are elaborated below.

10.8 Minimally Invasive and Robotic Surgery (MIS and MIRS)

MIS has revolutionized many surgical procedures over the last few decades. Early MIS, known as keyhole surgery, was performed using a small video camera, a video display, and a few customized surgical tools. Nowadays, much attention is given to tactile sensing in MIS. In procedures such as gall bladder removal (laparoscopic cholecystectomy), surgeons insert a camera and long slender tools into the abdomen through small skin incisions to explore the internal cavity and manipulate organs from outside the body as they view their actions on a video display. Because the development of minimally invasive techniques has reduced the sense of touch compared to open surgery, surgeons must rely more on the feeling of net forces resulting from tool–tissue interactions and so need more training to successfully operate on patients. Although tissue color and texture convey important anatomical information visually, touch is still critical in identifying otherwise obscure tissue planes, blood vessels, and abnormal tissues, and gauging optimal forces to be applied for tissue manipulation.

However, MIS still involves humans in the feedback loop and hence does not cover all needs for performing intelligent robotic manipulation. Outside the laboratory, manipulation is still primarily performed without tactile sensing. In an industrial setting, most variables can be controlled. Even though force/torque sensors are used for grinding operations and for peg-in-hole tasks, the really large benefits with a refined tactile sense can be reaped outside such well-controlled environments.

For a versatile robot in an uncertain environment, tactile sensing will open up new possibilities. Much of the art of MIS and training for a particular procedure depend upon the education and refinement of the trainee's haptic sensorimotor system. The benefits of using haptic devices in medical training through simulation [73, 109–114] have already been recognized by several research groups and many of the companies working in this area (for example, Immersion Medical, Surgical Science, Mentice, and Reachin Technologies). The rapid increase in the number of papers on haptics and surgical simulation published in international conference proceedings in the last five years indicates haptics' growing importance, although this rate of growth is not commensurate with the amount of research that has been undertaken in the realm of interface devices, algorithms, or even surgical training with and without haptics.

Telesurgical tasks require high dexterity, fidelity, and virtually authentic haptic feedback during manipulation since most of it is delicate so it therefore follows that the design requirements for teleoperation controllers need to be significantly different to those in classical teleoperation applications. An important component of the teleoperator design is the quantification of human operator sensitivity and performance.

Figures 10.9 and 10.10 show several telesurgical concepts in which the surgeon is physically separated from the workspace.

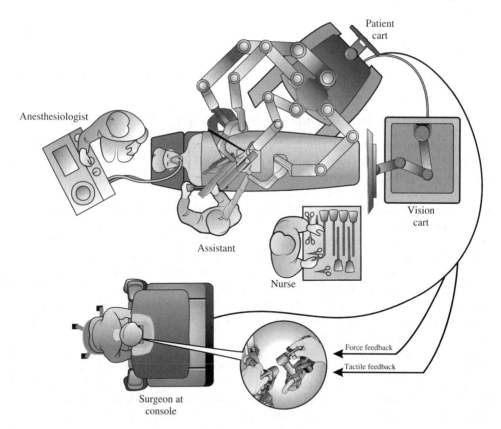

Figure 10.9   Telesurgical system concept. In this concept, surgeon is connected to the system by means of console which provides the surgeon with vision, and master controls which translate the surgeon's hand, wrist, and finger movements into precise, real-time movements of surgical instruments. The robotic arms in the patient cart are equipped with surgical instruments and let every surgical maneuver be under the direct control of the surgeon (See Plate 19)

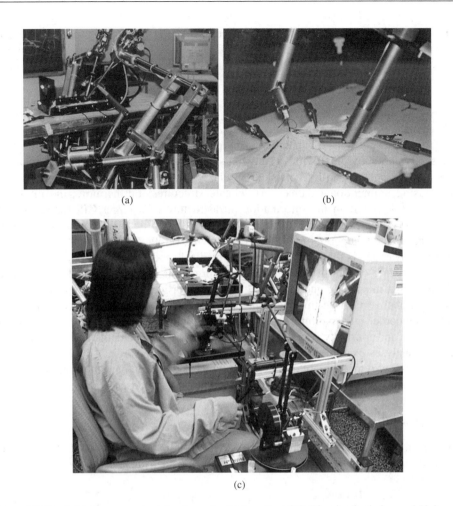

(a)   (b)

(c)

Figure 10.10 (a,b) Slave manipulator from the University of California, Berkeley and University of California, San Francisco laparoscopic telesurgical workstation, tying a knot in the training box. (c) Master workstation of the robotic telesurgical workstation (RTW) (See Plate 20)

The key advantages of medical robotics can be grouped into three areas. The first is the potential of a medical robot to significantly improve the surgeons' technical capability to perform procedures, by exploiting the complementary strengths of humans and robots. Medical robots can now be constructed to be more precise and geometrically accurate than an unaided human. They can operate in hostile radiological environments and can provide greater dexterity for minimally invasive procedures inside the patient's body. These capabilities can both enhance the ability of an average surgeon to perform procedures unassisted that otherwise only a few exceptionally gifted surgeons could perform. It also makes it possible to perform interventions that would otherwise be completely unfeasible, such as, for example, by using the da Vinci™ (Figure 9.1) Surgical System. This comparatively recent innovation simplifies many existing MIS and potentially expands upon the number of other such procedures than can be performed. It makes difficult operations routine and, in addition, these procedures are less traumatic for the patient and their outcome is greatly improved. Also, instruments used in the da Vinci® telemanipulator system have now been further modified for measuring forces while executing surgical tasks [115]. Since the shaft of the surgical instruments is made of carbon fiber, force sensors have to be very sensitive and reliable. Therefore strain gauge sensors, which are employed for industrial force registration, are applied at the distal end of the instrument's shaft, near the gripper, in order to display realistic forces during an operation.

A second, closely related capability is the potential of medical robots to promote surgical safety, both by improving a surgeon's technical performance and by means of active assists such as 'no-fly zones' or virtual fixtures to prevent surgical instruments from causing unintentional damage to delicate structures. Furthermore, the integration of medical robots within the information infrastructure of a larger CIS (computer integrated surgery) system can provide the surgeon with significantly improved monitoring and online decision supports, thus further improving safety.

A third advantage is the inherent ability of medical robots and CIS systems to promote consistency while capturing detailed online information for every procedure. Consistent execution (e.g., in spacing and tensioning of sutures or in placing of components in joint reconstructions) is itself an important quality factor. If saved and routinely analyzed, the flight data recorder information inherently available with a medical robot can be used both in morbidity and mortality assessments of serious surgical incidents and, potentially, in statistical analyses examining many cases to develop better surgical plans. Furthermore, such data can provide valuable input for surgical simulators, as well as a database for developing skill assessment and certification tools for surgeons.

10.9 Robotics

Many sensing devices have been developed for robotic manipulation. A sketch of a robot hand with some of the most common types of contact sensor is shown in Figure 10.11.

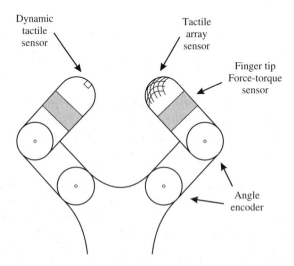

Figure 10.11 Schematic drawing of a robot hand equipped with several types of contact sensor [116]

These sensors are the tactile array sensor, fingertip force-torque sensor, and various dynamic tactile sensors.

At present, the types of contact information required for manipulation control have still to be fully established. Depending on the details of each task, the sensing requirements will certainly vary and will range from little or no sensing at all, as in the case of a simple pick-and-place operation in assembly operations, or a great deal of sensing and contact information as when rolling, sliding, or regrasping an unknown object during manipulation. One important distinction is between continuous sensing that is used in real-time control of the fingers, and simple threshold detection that is used in 'guarded moves.' There have been few reports of experimental investigations of manipulation with robot hands. Thus the sources of the hypotheses about sensing in manipulation presented here are mainly human contact sensing and robotic grasp analyses [115, 117, 118].

10.10 Virtual Environment

More and more research and corporate resources are being invested in the development of VEs. The challenge of virtual reality technology is to provide users with sensory stimulations that are as 'realistic' as possible, that is, for a given situation, producing a sensory flow giving rise to the same perception as that experienced in 'real' life. Most of the VEs built to date contain visual and spatialized sound displays of high fidelity, whereas haptic display technology that allows for manual interactions with these environments clearly lags behind. Yet, being able to touch, feel, and manipulate objects, in addition to seeing and hearing them, is essential for realizing the full promise of VEs. Indeed, haptic perception provides a sense of immersion in an environment that is otherwise not possible. In addition, one of the most important potential applications of virtual reality displays is the development of training and simulation systems, especially in the domains where real practice presents risks for the involved persons. In this regard, haptic displays have a critical role to play since the main part of our interactions with the environment involves tactual and manipulative skills. For instance, a haptic-based virtual reality display can be very useful in providing 'safe' surgery or medical exploration training to novice surgeons, in both the human and veterinary fields [119–122], in which any mistake can have life-threatening consequences. Unfortunately, even by employing the most advanced technological approach it is still not possible to construct a haptic device that generates a satisfactory facsimile of the real world, due to mechanical, temporal, and frequency constraints. In virtual reality environments, a person interacts with a simulated environment within which the person perceives virtual objects through the senses with a high degree of freedom. He/she can move objects, react to them, and manipulate them at their natural size as would be possible if they were real. For instance, monitors cannot display real movements of objects, but they are able to create an illusion of movement by successively displaying single pictures faster than human visual perception can resolve them. This example illustrates how taking into account the specificities of human perception, and exploiting its limits, can allow us to by-pass these technical limitations. Psychophysical experiments have traditionally used three methods for testing the subjects' perception in stimulus detection and difference detection, which include the *method of limits*, the *method of constant stimuli*, and the *method of adjustment*. In the method of limits, some property of the stimulus starts out at a level so low that it cannot be detected; then this level is gradually increased until the participant reports that they are aware of it. In the method of constant stimuli, the levels of a certain property of the stimulus are not related from one trial to the next, but presented randomly. Finally, the method of adjustment asks the subject to control the level of the stimulus, instructs them to alter it until it is just barely detectable against the background noise, or is the same as the level of another stimulus. All these requirements must be taken into account when developing virtual reality displays.

The haptic world consists primarily of objects, surfaces, and their properties, rather than of sensory inputs. Of course some of these environmental properties are directly coded by the sensory system and sometimes even just by a small subset of single receptors, such as with temperature or small discontinuities on a surface [123], but the perception of many of environmental properties is based on the integration of different sensory sources. These integration mechanisms bear further possibilities for the development of haptic technology, as they may allow for the substitution of one type of sensory input with another.

Whereas there has been a large body of theoretical and empirical research on visual space perception, there is no agreed-upon definition of haptic space. Distinctions have been made between manipulatory and ambulatory space, in which the former is within reach of the hands, whereas the latter requires exploration by movements of the body. Both involve haptic feedback, although to different effectors. A variety of studies have established that the perception of manipulatory space is nonveridical although there have been studies that have demonstrated various distortions in space perception which is, itself, anisotropic. People, even those who are congenitally blind, regularly overestimate the length of touched vertical lines as compared to horizontal ones [124, 125]. They also regularly overestimate the length of radial movements (to and from the body) by about 10% compared to tangential movements [126]. As a further example, the orientation of a line that is felt in space is greatly dependent on its position (especially tangential) with respect to the body. Furthermore, when attempting to orient bars in parallel by means of touch alone, deviations of up to 40% can result, especially if done bimanually [92, 93, 127]. Therefore, obliquely oriented bars are less accurately reproduced compared to those oriented in line with the axis of the body [128, 129]. Many further distortions in the form of optical illusions have been reported, such as that given by Mueller-Lyer [130]. Attempts to describe these different distortions by some kind of nonveridical, but inherently consistent haptic metrics, proved unsuccessful. For example, a metric derived from distortions in perceived angles at different positions in space did not fit with one derived from perceived length [131].

However, there are numerous studies relating these spatial distortions to different factors within the movements. For example, body and arm position play a role in the size of horizontal–vertical and radial–tangential illusions [132, 133]. The latter illusions were accounted for by different movement velocities in the radial and tangential directions and vanished when people were obliged to assimilate the velocities of their movements in the two different directions [134]. This corresponds to the observation that movement velocity strongly affects estimates of the length of a line up to a factor of three (reported for velocities between 0.5 to 50 cm s^{-1} [133]). Furthermore, oblique effects (in the vertical plane parallel to the observer) seem to relate to gravitational forces [128], at least in part [129].

There are other strong illusions directly related to movement patterns. If people move along a curved line and then attempt to estimate the shortest distance between start and end point, the longer the movement path, the more the distance is overestimated (up to a factor of two). This is true even if one hand remains on the starting point [135, 136]. Furthermore, if people move one arm against a constant force, the other arm can estimate its position appropriately, but if the force varies during the movement people make systematic mistakes [137–139]. This holds also true for actively induced force changes, for example, when matching the indentation depth of two springs with different compliances [140].

Altogether, the number of illusions in the perception of manipulatory haptic space is large and may, especially, depend upon the way the kinesthetic system derives position and movement of limbs from the muscle receptors. As noted above, this relationship is not presently well understood. Thus, although the illusions are rather interesting for basic research, it is still not easy to derive general rules and apply them to haptic display design. Consequently, the following suggestions remain highly speculative and will need to be subjected to further investigation.

One general rule may be derived from the fact that in everyday life, when we touch and look around in space, we are not aware of all these distortions. Visual distortions in the same direction can explain this fact to a small extent [141]. Primarily, it may mean that with vision available we do not care much about coarse haptic relationships and, consequently, it may be of minor importance to get these exactly right in the haptic part of virtual reality. Moreover, specific distortions may suggest specific and probably useful applications or directions of research. For example, virtual realities that visually stretch empty space in haptically overestimated directions may be able to increase virtual haptical workspace somewhat, albeit unnoticed by the user. Or, the fact that space perception by the two different hands is especially inconsistent could promise simplification in the synchronization of bimanual interfaces.

Finally, space distortions stemming from varying forces have been shown to have a promising counterpart in shape perception, which will be discussed.

In VEs, the perception of objects (and surfaces) in haptic space can be split into the perception of their different haptic properties, such as material and geometry, as in real haptic systems. Compared to the other senses, haptics is the only one that can determine the material property of an object, such as its weight and surface properties which, itself, has been further differentiated into perceptual categories such as roughness, softness, stickiness, and apparent temperature [142], and through which most people primarily tend to make their determination and formulate their perception of softness and stickiness [143]. This may mean that the perception resulting from a surface is well described in terms of these two or three dimensions. Likewise the display of haptic material in VEs probably will be limited but should, in any case, concentrate on these two or three dimensions.

References

1. Fearing, R.S., Moy, G., and Tan, E. (1997) Some basic issues in teletaction. Proceedings of the 1997 IEEE International Conference on Robotics and Automation, 1997, vol. 4, pp. 3093–3099.
2. Dargahi, J. (2000) A piezoelectric tactile sensor with three sensing elements for robotic, endoscopic and prosthetic applications. *Sensors and Actuators A: Physical*, **80**, 23–30.
3. Mirbagheri, A., Dargahi, J., Najarian, S., and Ghomshe, F.T. (2007) Design, fabrication, and testing of a membrane Piezoelectric tactile sensor with four sensing elements. *American Journal of Applied Sciences*, **4**, 645–652.

4. Geisthoff, U.W., Tretbar, S.H., Federspil, P.A., and Plinkert, P.K. (2004) Improved ultrasound-based navigation for robotic drilling at the lateral skull base. *International Congress Series*, **1268**, 662–666.
5. Dargahi, J. (2002) An endoscopic and robotic tooth-like compliance and roughness tactile sensor. *Journal of Mechanical Engineering Design (ASME)*, **124**, 576–582.
6. Emamieh, G.D., Ameri, A., Najarian, S., and Golpaygani, A.T. (2008) Experimental and theoretical analysis of a novel flexible membrane tactile sensor. *American Journal of Applied Sciences*, **5**, 122–128.
7. Ly, U.-L., Bryson, A.E., and Cannon, R.H. (1985) Design of low-order compensators using parameter optimization. *Automatica*, **21**, 315–318.
8. Massimino, M.J. and Sheridan, T.B. (1993) Sensory substitution for force feedback in teleoperation. *Presence: Teleoperators and Virtual Environments*, **2**, 344–352.
9. Patrick, N.J.M., Sheridan, T.B., Massimino, M.J., and Marcus, B.A. (1991) Design and testing of a nonreactive, fingertip, tactile display for interaction with remote environments. *Proceedings of SPIE*, **1387**, 215–222.
10. Akamatsu, M. (1994) Touch with a mouse-a mouse type interface device with tactile and force display. Proceedings of the 3rd IEEE International Workshop on Robot and Human Communication, 1994. RO-MAN '94 Nagoya, 1994, pp. 140–144.
11. Howe, R.D., Peine, W.J., Kantarinis, D.A., and Son, J.S. (1995) Remote palpation technology. *IEEE Engineering in Medicine and Biology Magazine*, **14**, 318–323.
12. Howe, R.D., Kontarinis, D.A. and Dimitrios, A.(1993) Task performance with a dexterous teleoperated hand system. *Proceedings of SPIE*, **1833**, 199–207.
13. Howe, R.D. and Kontarinis, D.A. (1994) High-frequency force information in teleoperated manipulation. Presented at the The 3rd International Symposium on Experimental Robotics III, 1994.
14. Bejczy, A.K. (1980) Sensors, controls, and man–machine interface for advanced teleoperation. *Science*, **208**, 1327–1335.
15. Hirzinger, G., Brunner, B., Dietrich, J., and Heindl, J. (1993) Sensor-based space robotics-ROTEX and its telerobotic features. *IEEE Transactions on Robotics and Automation*, **9**, 649–663.
16. Cavusoglu, M.C., Tendick, F., Cohn, M., and Sastry, S.S. (1999) A laparoscopic telesurgical workstation. *IEEE Transactions on Robotics and Automation*, **15**, 728–739.
17. Liu, A., Tharp, G., French, L. *et al.* (1993) Some of what one needs to know about using head-mounted displays to improve teleoperator performance. *IEEE Transactions on Robotics and Automation*, **9**, 638–648.
18. Hagner, D. and Webster, J.G. (1988) Telepresence for touch and proprioception in teleoperator systems. *IEEE Transactions on Systems, Man and Cybernetics*, **18**, 1020–1023.
19. Bicchi, A., Scilingo, E.P., and De Rossi, D. (2000) Haptic discrimination of softness in teleoperation: the role of the contact area spread rate. *IEEE Transactions on Robotics and Automation*, **16**, 496–504.
20. Itoh, T., Kosuge, K., and Fukuda, T. (2000) Human-machine cooperative telemanipulation with motion and force scaling using task-oriented virtual tool dynamics. *IEEE Transactions on Robotics and Automation*, **16**, 505–516.
21. Zhang, H. and Chen, N.N. (2000) Control of contact via tactile sensing. *IEEE Transactions on Robotics and Automation*, **16**, 482–495.
22. CyberGlove Systems LLC CyberGlove. www.cyberglovesystems.com, accessed 2012.
23. Walker, M. (2007) Video Conferencing: A Guide to Making a Telepresence Business Case, www.thevideonetwork.co.uk/accessed 2012.
24. Edmondson, R.S. (2005) Evaluating the effectiveness of a telepresence-enabled cognitive apprenticeship model for teacher professional development. PhD, Utah State University, Logan, UT.
25. Kahled, W., Reichling, S., Bruhns, O.T. *et al.* (2004) Palpation imaging using a haptic system for virtual reality applications in medicine. Proceedings of the 12th Annual Medicine Meets Virtual Reality Conference, Newport Beach, CA, pp. 147–153.
26. Moy, G. and Singh, U. (2000) Human psychophysics for teletaction system design. *The Electronic Journal of Haptics Research*, **1**.
27. Shimoga, K.B. (1992) Finger force and touch feedback issues in dexterous telemanipulation. Proceedings of the 4th Annual Conference on Intelligent Robotic Systems for Space Exploration, 1992, pp. 159–178.
28. Fechner, G.T. (1860) Specielles zur Methode richtiger und falscher Fälle, in *Elemente der Psychophysik*, vol. **1**, Breitkopf und Härte, Leipzig, pp. 93–120.
29. Fancher, R.E. (1996) *Pioneers of Psychology*, 3rd edn, W. W. Norton & Company, New York.
30. Sheynin, O. (2004) Fechner as a statistician. *The British Journal of Mathematical and Statistical Psychology*, **57** (Pt 1), 53–72.

31. Kaernbach, C., Schroger, E., and Muller, H. (2004) *'Psychophysics Beyond Sensation' Laws and Invariants of Human Cognition*, Publication: Lawrence Erlbaum Associates, Mahwah, NJ.

32. Zwislocki, J.J. (2009) *Sensory Neuroscience: Four Laws of Psychophysics*, Springer.

33. Makila, P. and Toivonen, H. (1987) Computational methods for parametric LQ problems--A survey. *IEEE Transactions on Automatic Control*, **32**, 658–671.

34. Stevens, S.S. (1957) On the psychophysical law. *Psychological Review*, **64**, 153–181.

35. Zwislocki, J. (1983) Group and individual relations between sensation magnitudes and their numerical estimates. *Attention, Perception & Psychophysics*, **33**, 460–468.

36. Hellman, R.P. and Zwislocki, J. (1961) Some factors affecting the estimation of loudness. *The Journal of the Acoustical Society of America*, **33**, 687–694.

37. Hellman, R.P. and Zwislocki, J. (1963) Monaural loudness function at 1000 cps and interaural summation. *The Journal of the Acoustical Society of America*, **35**, 856–865.

38. Robinson, D.W. (1957) The subjective loudness scale. *Acustica*, **7**, 217–233.

39. Stevens, S.S. (1956) The direct estimation of sensory magnitudes: loudness. *The American Journal of Psychology*, **69**, 1–25.

40. Scharf, B. and Stevens, J.C. (1961) The form of the loudness function near threshold. Proceedings of the 3rd International Congress on Acoustics, Amsterdam, 1961, pp. 80–82.

41. Feldtkeller, R., Zwicker, E., and Port, E. (1959) Lautstärke, Verhältnislautheit und Summenlautheit. *Frequenz*, **13**, 108–117.

42. Phillips, J.R. and Johnson, K.O. (1981) Tactile spatial resolution. II. Neural representation of Bars, edges, and gratings in monkey primary afferents. *Journal of Neurophysiology*, **46**, 1192–1203.

43. Phillips, J.R. and Johnson, K.O. (1981) Tactile spatial resolution. III. A continuum mechanics model of skin predicting mechanoreceptor responses to bars, edges, and gratings. *Journal of Neurophysiology*, **46**, 1204–1225.

44. Loomis, J.M. and Lederman, S.J. (1986) Tactual perception, in *Handbook of Perception and Human Performance* (eds K.R. Boff, L. Kaufman, and J.P. Thomas), John Wiley & Sons, Inc., New York, pp. 31/1–31/41.

45. Shimojo, M., Shinohara, M., and Fukui, Y. (1999) Human shape recognition performance for 3D tactile display. *IEEE Transactions on Systems, Man and Cybernetics, Part A: Systems and Humans*, **29**, 637–644.

46. Johansson, R.S. and Vallbo, A.B. (1979) Detection of tactile stimuli. Thresholds of afferent units related to psychophysical thresholds in the human hand. *The Journal of Physiology*, **297**, 405–422.

47. LaMotte, R.H. and Srinivasan, M.A. (1987) Tactile discrimination of shape: responses of rapidly adapting mechanoreceptive afferents to a step stroked across the monkey fingerpad. *The Journal of Neuroscience*, **7**, 1672–1681.

48. Johansson, R.S., Landström, U., and Lundström, R. (1982) Responses of mechanoreceptive afferent units in the glabrous skin of the human hand to sinusoidal skin displacements. *Brain Research*, **244**, 17–25.

49. Johnson, K.O. and Phillips, J.R. (1981) Tactile spatial resolution. I. Two-point discrimination, gap detection, grating resolution, and letter recognition. *Journal of Neurophysiology*, **46**, 1177–1192.

50. Lederman, S.J. (1987) Heightening tactile impressions of surface texture, in *Active Touch* (ed. G. Gordon), Pergamon Press, Oxford, pp. 205–214.

51. Gordon, I.E. and Cooper, C. (1975) Improving one's touch. *Nature*, **256**, 203–204.

52. Maeno, T. and Kobayashi, K. (1998) FE analysis of the dynamic characteristics of the human finger pad in contact with objects with/without surface roughness. IMECE Proceedings of the ASME Dynamic Systems and Controls Division, Anaheim, CA, pp. 279–286.

53. Enns, J.T. (2004) *The Thinking Eye, The Seeing Brain: Explorations in Visual Cognition*, W. W. Norton & Company, New York.

54. Hatwell, Y. and Streri, A. (2003) *Touching for Knowing*, John Benjamins Publishing Company.

55. Humphreys, G.W., Price, C.J., and Riddoch, M.J. (1999) From objects to names: a cognitive neuroscience approach. *Psychological Research*, **62**, 118–130.

56. Riddoch, M.J. and Humphreys, G.W. (2001) Object recognition, in *Handbook of Cognitive Neuropsychology* (ed. B. Rapp), Psychology Press, Hove.

57. Ward, J. (2006) *The Student's Guide to Cognitive Neuroscience*, Psychology Press, New York.

58. Bar, M. (2003) A cortical mechanism for triggering top-down facilitation in visual object recognition. *Journal of Cognitive Neuroscience*, **15**, 600–609.

59. Marr, D. (1976) Early processing of visual information. *Philosophical Transactions of the Royal Society of London B, Biological Sciences*, **275**, 483–519.

60. Richer, F. and Boulet, C. (1999) Frontal lesions and fluctuations in response preparation. *Brain and Cognition*, **40**, 234–238.
61. Humphreys, G.W., Quinlan, P.T., and Riddoch, M.J. (1987) Normal and pathological processes in visual object constancy, in *Visual Object Processing: A Cognitive Neuropsychological Approach* (ed. G.W. Humphreys and M.J. Riddoch), Lawrence Erlbaum Associates, Hove, UK, 43–105.
62. Peterson, M.A. and Rhodes, G. (2003) *Perception of Faces, Objects and Scenes: Analytic and Holistic Processes*, Oxford University Press, New York.
63. Tarr, M.J. and Bülthoff, H.H. (1995) Is human object recognition better described by geon structural descriptions or by multiple views? Comment on Biederman and Gerhardstein (1993). *Journal of Experimental Psychology: Human Perception and Performance*, **21**, 1494–1505.
64. Marr, D. and Nishihara, H.K. (1978) Representation and recognition of the spatial organization of three-dimensional shapes. *Proceedings of the Royal Society of London, Series B: Biological Sciences*, **200**, 269–294.
65. Biederman, I. (1987) Recognition by components: a theory of human image understanding. *Psychological Review*, **94**, 115–147.
66. Loomis, J.M. (1990) A model of character recognition and legibility. *Journal of Experimental Psychology: Human Perception and Performance*, **16**, 106–120.
67. Gibson, J.J. (1962) Observations on active touch. *Psychological Review*, **69**, 477–491.
68. Gibson, J.J. (1966) *The Senses Considered as Perceptual Systems*, Houghton Mifflin, Boston.
69. Okamura, A.M. and Cutkosky, M.R. (2001) Feature detection for haptic exploration with robotic fingers. *The International Journal of Robotics Research*, **20**, 925–938.
70. Lederman, S.J. and Klatzky, R.L. (1990) Haptic classification of common objects: knowledge-driven exploration. *Cognitive Psychology*, **22**, 421–459.
71. Lederman, S.J. and Klatzky, R.L. (1997) Relative availability of surface and object properties during early haptic processing. *Journal of Experimental Psychology: Human Perception and Performance*, **23**, 1680–1707.
72. Treisman, A. and Gormican, S. (1988) Feature analysis in early vision: evidence from search asymmetries. *Psychological Review*, **95**, 15–48.
73. Székely, G., Brechbühler, C., Hutter, R. *et al.* (1998) Modeling of soft tissue deformation for laparoscopic surgery simulation. Proceedings of the 1st International Conference on Medical Image Computing and Computer-Assisted Intervention (MICCAI), LNCS 1496, pp. 550–561.
74. Pick, A.D. and Pick, H.L. (1966) A developmental study of tactual discrimination in blind and sighted children and adults. *Psychonomic Science*, **6**, 367–368.
75. Brumaghin, S.H. and Brown, D.R. (1969) Perceptual equivalence between visual and tactual stimuli: an anchoring study. *Perception and Psychophysics*, **4**, 175–179.
76. Owen, D. and Brown, D. (1970) Visual and tactual form discrimination: A psychophysical comparison within and between modalities. *Perception and Psychophysics*, **7**, 302–306.
77. Garbin, C. and Bernstein, I. (1984) Visual and haptic perception of three-dimensional solid forms. *Attention, Perception & Psychophysics*, **36**, 104–110.
78. Garbin, C. (1990) Visual-touch perceptual equivalence for shape information in children and adults. *Attention, Perception & Psychophysics*, **48**, 271–279.
79. Hatwell, Y., Orliaguet, J.P., and Brouty, G. (1990) Effects of object properties, attentional constraints and manual exploratory procedures on haptic perceptual organization: a developmental study, in *Sensory-Motor Organizations and Development in Infancy and Early Childhood* (eds H. Bloch and B. Bertenthal), Klumer Academic Publishers, Dordrecht, pp. 315–335.
80. Lakatos, S. and Marks, L. (1999) Haptic form perception: relative salience of local and global features. *Attention, Perception & Psychophysics*, **61**, 895–908.
81. Locher, P.J. and Wagemans, J. (1993) Effects of element type and spatial grouping on symmetry detection. *Perception*, **22**, 565–587.
82. Wagemans, J. (1995) Detection of visual symmetries. *Spatial Vision*, **9**, 9–32.
83. Walk, R.D. (1965) Tactual and visual learning of forms differing in degree of symmetry. *Psychonomic Science*, **2**, 93–94.
84. Ballesteros, S., Manga, D., and Reales, J. (1997) Haptic discrimination of bilateral symmetry in 2-dimensional and 3-dimensional unfamiliar displays. *Attention, Perception & Psychophysics*, **59**, 37–50.
85. Ballesteros, S., Millar, S., and Reales, J. (1998) Symmetry in haptic and in visual shape perception. *Attention, Perception, & Psychophysics*, **60**, 389–404.

86. Millar, S. (1994) *Understanding and Representing Space. Theory and Evidence from Studies with Blind and Sighted Children*, Clarendon Press, Oxford.
87. Appelle, S. (1972) Perception and discrimination as a function of stimulus orientation: the 'oblique effect' in man and animals. *Psychological Bulletin*, **78**, 266–278.
88. Gentaz, E., Luyat, M., Cian, C. *et al*. (2001) The reproduction of vertical and oblique orientations in the visual, haptic, and somatovestibular systems. *The Quarterly Journal of Experimental Psychology Section A*, **54**, 513–526.
89. Ballaz, C. and Gentaz, E. (2000) La perception visuelle des orientations et l'effet de l'oblique. *L'année Psychologique*, **100**, 715–744.
90. Gentaz, E. and Tshopp, C. (2002) The oblique effect in the visual perception of orientations, in *Advances in Psychology Research* (ed. P. Shovoh), Nova Sciences Publishers, New York, pp. 137–163.
91. Kappers, A.M.L. (1999) Large systematic deviations in the haptic perception of parallelity. *Perception*, **28**, 1001–1012.
92. Kappers, A.M.L. (2002) Haptic perception of parallelity in the midsagittal plane. *Acta Psychologica*, **109**, 25–40.
93. Kappers, A.M.L. and Koenderink, J.J. (1999) Haptic perception of spatial relations. *Perception*, **28**, 781–795.
94. Cuijpers, R.H., Kappers, A.M.L., and Koenderink, J.J. (2000) Large systematic deviations in visual parallelism. *Perception*, **29**, 1467–1482.
95. Cuijpers, R.H., Kappers, A.M.L., and Koenderink, J.J. (2002) Visual perception of collinearity. *Perception & Psychophysics*, **64**, 392–404.
96. Hatwell, Y. (1986) *Toucher L'espace: La Main et la Perception Tactile de L'espace*, Presses universitaires de Lille, Lille.
97. Phillips, J.R. and Johnson, K.O. (1981) Tactile spatial resolution. II. Neural representation of bars, edges, and gratings in monkey primary afferents. *Journal of Neurophysiology*, **46**, 1192–1203.
98. Anderson, R.J. and Spong, M.W. (1989) Bilateral control of teleoperators with time delay. *IEEE Transactions on Automatic Control*, **34**, 494–501.
99. Hannaford, B. (1989) A design framework for teleoperators with kinesthetic feedback. *IEEE Transactions on Robotics and Automation*, **5**, 426–434.
100. Lawrence, D.A. (1993) Stability and transparency in bilateral teleoperation. *IEEE Transactions on Robotics and Automation*, **9**, 624–637.
101. Yokokohji, Y. and Yoshikawa, T. (1994) Bilateral control of master–slave manipulators for ideal kinesthetic coupling-formulation and experiment. *IEEE Transactions on Robotics and Automation*, **10**, 605–620.
102. Hajian, A.Z. and Howe, R.D. (1994) Identification of the mechanical impedance at the human fingertip. ASME International Mechanical Engineering Congress and Exposition Chicago, IL, pp. 319–327.
103. Takei, K., Takahashi, T., Ho, J.C. *et al*. (2010) Nanowire active-matrix circuitry for low-voltage macroscale artificial skin. *Nature Materials*, **9**, 821–826.
104. Fearing, R.S., Moy, G., and Tan, E. (1997) Some basic issues in teletaction. Proceedings of the IEEE International Conference on Robotics and Automation, 1997, vol. 4, pp. 3093–3099.
105. Nicolson, E.J. and Fearing, R.S. (1993) Sensing capabilities of linear elastic cylindrical fingers. Proceedings of the 1993 IEEE/RSJ International Conference on Intelligent Robots and Systems '93, IROS '93, vol. 1, pp. 178–185.
106. Sheridan, T.B. and Ferrell, W.R. (1963) Remote manipulative control with transmission delay. *IEEE Transactions on Human Factors in Electronics*, **HFE-4**, 25–29.
107. Kim, J., Kim, H., Tay, B.K. *et al*. (2004) Transatlantic touch: a study of haptic collaboration over long distance. *Presence Teleoper Virtual Environments*, **13**, 328–337.
108. Robles-De-La-Torre, G. (2009) Virtual reality: touch/haptics, in *Encyclopedia of Perception*, vol. **2** (ed. B. Goldstein), Sage Publication, Thousand Oaks, CA, pp. 1036–1038.
109. Basdogan, C., Ho, C.H., and Srinivasan, M.A. (2001) Virtual environments for medical training: graphical and haptic simulation of laparoscopic common bile duct exploration. *IEEE/ASME Transactions on Mechatronics*, **6**, 269–285.
110. Cakmak, H.K. and Kühnapfel, U. (2000) Animation and simulation techniques for VR-training systems in endoscopic surgery. Proceedings of the 11th Eurographics Workshop on Endoscopic Surgery, Computer Animation and Simulation 2000, pp. 173–185.
111. Cotin, S., Delingette, H., and Ayache, N. (1999) Real-time elastic deformations of soft tissues for surgery simulation. *IEEE Transactions on Visualization and Computer Graphics*, **5**, 62–73.

112. Brown, J., Sorkin, S., Latombe, J-C. *et al*. (2002) Algorithmic tools for real-time microsurgery simulation. *Medical Image Analysis*, **6**, 289–300.

113. Wu, X., Downes, M.S., Goktekin, T., and Tendick, F. (2001) Adaptive nonlinear finite elements for deformable body simulation using dynamic progressive meshes. *Computer Graphics Forum*, **20**, 349–358.

114. O'Toole, R.V., Playter, R.R., Krummel, T.M. *et al*. (1999) Measuring and developing suturing technique with a virtual reality surgical simulator. *Journal of the American College of Surgeons*, **189**, 114–127.

115. Salisbury, J. Jr., (1984) Interpretation of contact geometries from force measurements. Proceedings of the 1984 IEEE International Conference on Robotics and Automation, 1984, pp. 240–247.

116. Howe, R. (1994) Tactile sensing and control of robotic manipulation. *Journal of Advanced Robotics*, **8**, 245–261.

117. Kerr, J. and Roth, B. (1986) Analysis of Multifingered Hands. *The International Journal of Robotics Research*, **4**, 3–17.

118. Cole, A.A., Hsu, P., and Sastry, S.S. (1989) Dynamic regrasping by coordinated control of sliding for a multifingered hand. Proceedings of the 1989 IEEE International Conference on Robotics and Automation, 1989, vol. 2, pp. 781–786.

119. Sherman, K.P., Ward, J.W., Wills, D.P., and Mohsen, A.M. (1999) A portable virtual environment knee arthroscopy training system with objective scoring. *Studies in Health Technology and Informatics*, **62**, 335–336.

120. Kühnapfel, U., Çakmak, H.K., and Maaß, H. (1999) 3D Modeling for endoscopic surgery. IEEE Symposium on Simulation, Delft, NL, 1999, pp. 22–32.

121. Bro-Nielsen, M., Tasto, J.L., Cunningham, R., and Merril, G.L. (1999) PreOp endoscopic simulator: a PC-based immersive training system for bronchoscopy. *Studies in Health Technology and Informatics*, **62**, 76–82.

122. Delp, S.L., Loan, P., Basdogan, C., and Rosen, J.M. (1997) Surgical simulation: an emerging technology for training in emergency medicine. *Presence*, **6**, 147–159.

123. Vierck, C.J.J. (1979) Comparisons of punctate, edge and surface stimulation of peripheral, slowly-adapting, cutaneous, afferent units of cats. *Brain Research*, **175**, 155–159.

124. Casla, M., Blanco, F., and Travieso, D. (1999) Haptic perception of geometric illusions by persons who are totally congenitally blind. *Journal of Visual Impairment & Blindness*, **93**, 583–588.

125. Heller, M., Calcaterra, J., Burson, L., and Green, S. (1997) The tactual horizontal-vertical illusion depends on radial motion of the entire arm. *Attention, Perception, & Psychophysics*, **59**, 1297–1311.

126. Cheng, M.-F. (1968) Tactile-kinesthetic perception of length. *The American Journal of Psychology*, **81**, 74–82.

127. Blumenfeld, W. (1937) The relationship between the optical and haptic construction of space. *Acta Psychologica*, **2**, 125–174.

128. Gentaz, E. and Hatwell, Y. (1995) The haptic 'oblique effect' in children's and adults' perception of orientation. *Perception*, **24**, 631–646.

129. Luyat, M., Gentaz, E., Corte, T.R., and Guerraz, M. (2001) Reference frames and haptic perception of orientation: body and head tilt effects on the oblique effect. *Perception Psychophisics*, **63**, 541–554.

130. Heller, M.A., Brackett, D.D., Wilson, K. *et al*. (2002) The haptic Müller-Lyer illusion in sighted and blind people. *Perception*, **31**, 1263–1274.

131. Fasse, E.D., Hogan, N., Kay, B.A., and Mussa-Ivaldi, F.A. (2000) Haptic interaction with virtual objects. spatial perception and motor control. *Biological Cybernetics*, **82**, 69–83.

132. Millar, S. and al Attar, Z. (2000) Vertical and bisection bias in active touch. *Perception*, **29**, 481–500.

133. Wong, T.S. (1977) Dynamic properties of radial and tangential movements as determinants of the haptic horizontal-vertical illusion with an 'L' figure. *Journal of Experimental Psychology: Human Perception and Performance*, **3**, 151–164.

134. Armstrong, L. and Marks, L.E. (1999) Haptic perception of linear extent. *Percept Psychophys*, **61**, 1211–1226.

135. Lederman, S.J., Klatzky, R.L., and Barber, P.O. (1985) Spatial and movement-based heuristics for encoding pattern information through touch. *Journal of Experimental Psychology: General*, **114**, 33–49.

136. Lederman, S.J. and Klatzky, R.L. (1987) Hand movements: a window into haptic object recognition. *Cognit Psychol*, **19**, 342–368.

137. Matthews, P. (1982) Where does sherrington's muscular sense originate? muscles, joints, corollary discharges? *Annual Review of Neuroscience*, **5**, 189–218.

138. Roland, P.E. (1977) A quantitative analysis of sensations of tension and of kinaesthesia in man. Evidence for a peripherally originating muscular sense and for a sense of effort. *Brain*, **100**, 671–692.
139. Rymer, W.Z. and D'Almeida, A. (1980) Joint position sense: the effects of muscle contraction. *Brain*, **103**, 1–22.
140. Roland, P.E. (1978) Sensory feedback to the cerebral cortex during voluntary movement in man. *Behavioral and Brain Sciences*, **1**, 129–171.
141. Bingham, G.P., Zaal, F., Robin, D., and Shull, J.A. (2000) Distortions in definite distance and shape perception as measured by reaching without and with haptic feedback. *Journal of Experimental Psychology-human Perception and Performance*, **26**, 1436–1460.
142. Klatzky, R.L. and Lederman, S.J. (1993) Toward a computational model of constraint-driven exploration and haptic object identification. *Perception*, **22**, 597–621.
143. Hollins, M., Bensmaïa, S., Karlof, K., and Young, F. (2000) Individual differences in perceptual space for tactile textures: Evidence from multidimensional scaling. *Attention, Perception & Psychophysics*, **62**, 1534–1544.

11

Teletaction Using a Linear Actuator Feedback-Based Tactile Display

In previous chapters, different techniques were discussed for developing tactile displays in teletaction applications. For example, Chapter 8 presented a graphical tactile display in which the patient could view tactile information rather than actually sensing it.

In this chapter, a tactile display is introduced which is able to reconstruct the softness of different materials based on their inherent characteristic properties. From an analogous point of view, most materials can be modeled as if they were springs. The compressibility of an ideal spring is linear and, indeed, remains so when the compression is small; within this range, the softness of a material can be characterized with accuracy. When measuring greater compression, however, the spring becomes more compressed and enters a range in which it becomes proportionately more nonlinear, and affects measurement accuracy. Most tactile displays that are driven by a servomotor, or any kind of actuator, use the spring model to simulate softness [1]. In this research work, the nonlinear behavior of materials is considered to describe the softness of touched materials. The objective of this work is to provide the user with a tactile display that is able to simulate different material compliances.

11.1 System Design

In this design, an actuator which converts rotational motion of a motor to linear motion was used as a softness display. The idea of this was to measure the force applied by the finger to a linear actuator shaft, then calculate the displacement of the shaft in response to the applied force, according to the mechanical properties of the simulated material, and move the shaft to the calculated position. In this study, one actuator shaft and one

Tactile Sensing and Displays: Haptic Feedback for Minimally Invasive Surgery and Robotics, First Edition.
Saeed Sokhanvar, Javad Dargahi, Siamak Najarian, and Siamak Arbatani.
© 2013 John Wiley & Sons, Ltd. Published 2013 by John Wiley & Sons, Ltd.

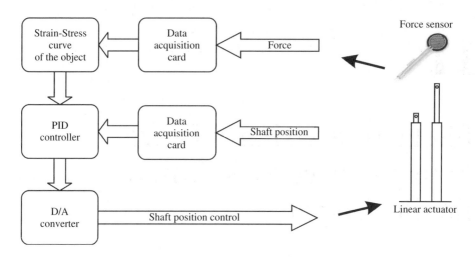

Figure 11.1 Block diagram of the tactile display

sensor was used to show the feasibility of the approach and this idea can be expanded to a matrix of pins utilizing miniaturization techniques.

A force sensor is used to measure the force exerted by a finger to the tip of the shaft. The measured force is registered by a data acquisition card and interpreted by processing software. The position of the shaft is also measured and transmitted to this same processing software through a data acquisition card. The mechanical properties of the object to be simulated are, in this case, the force–displacement data analogous to the stress–strain relations for that object, which have already been determined and stored. Knowing the force applied to the linear actuator by the fingertip, and the dimensions of the cap over which the force is applied, the stress can be calculated and the desired strain can be found by using the characteristic curve of the simulated object. The desired strain is then converted into a displacement and is used as the input for a proportional integral derivative (PID) controller, which is a generic control loop feedback device widely used in industrial control systems.

Having obtained the actual and desired position of the shaft, the PID prepares the control commands for the actuator and transmits them to the driver circuit. Figure 11.1 shows the block diagram of the system. A photograph of the tactile display is shown in Figure 11.2.

11.2 Tactile Actuator

The tactile display, shown in Figure 11.3, consists of a linear actuator, a Plexiglas cap (glued to the tip of shaft), a force sensor, a shaft position sensor and current driver electronics. The shaft of the actuator is able to move up to 2 cm in the vertical direction and can apply forces of up to 30 N. A force sensor, placed on the cap, measured the force applied by a finger to the shaft and a built-in position sensor was used to measure the

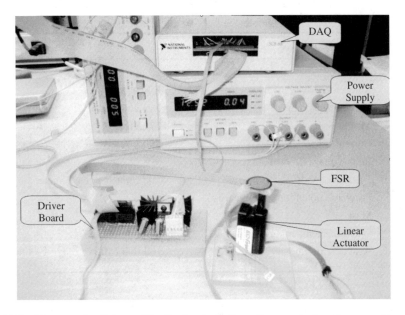

Figure 11.2   Photograph of the tactile display including: linear actuator, force-sensitive resistor (FSR), data acquisition card (DAQ), and driver circuit (See Plate 21)

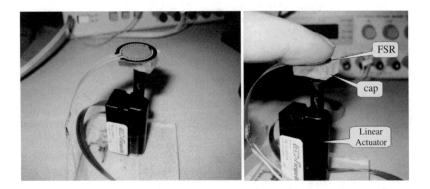

Figure 11.3   The linear actuator, FSR, and Plexiglas cap (See Plate 22)

position of the shaft with an error of less than 0.5 mm. The specifications of the linear actuator from Firgelli Technologies Inc., are presented in Table 11.1.

11.3   Force Sensor

A force-sensing resistor (FSR) is a device that consists of a polymer thick film (PTF) the resistance of which decreases in proportion to the force applied to its active surface. This device is extensively used in human touch control of electronic devices. Figure 11.4 demonstrates the structure of the sensor and its different components.

Table 11.1 The linear actuator specifications

Specifications	PQ-12
Stroke	20 mm
Rated force	6 N at 15 mm s^{-1}
Max power point	7 N at 13 mm s^{-1}
Max speed	27 mm s^{-1}
Current draw	250 mA
Input voltage	5 V dc
Feedback mode	2 kΩ linear potentiometer
Feedback potentiometer linearity	1%
Mass	75 g
Operating temperature	-10 to $+50$ °C
Actuator lifetime	100 000 cycles

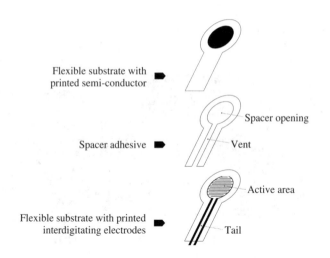

Figure 11.4 The structure of the sensor and its different components

To experimentally determine the force–conductance relationship for the FSR, several standard weights are placed on the active area and its conductance is measured. The conductance versus force data is plotted in Figure 11.5. The data plotted in Figure 11.5 show that the conductance in FSR is almost linearly related to the applied force.

Fitting a line on the experimental data, the following equation can be used to relate the conductance (C) and force (F).

$$C = a.F + b \tag{11.1}$$

Best fit using least squares method resulted in values of $a = 0.079$ and $b = -0.016$.

To find the resistance of the FSR, we used the electric circuit shown in Figure 11.6.

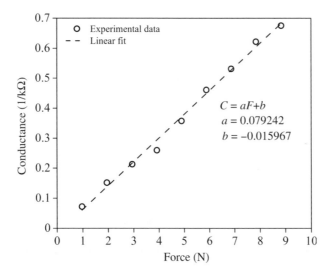

Figure 11.5 The conductance of the FSR versus force

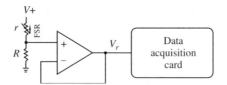

Figure 11.6 The electric circuit for measuring the resistance 'r'

In this figure the conductance of the FSR can be calculated from Equation 11.2.

$$V_r = \frac{R}{R+r} V^+ \Rightarrow C = \frac{1}{r} = \frac{1}{\left(\frac{V^+}{V_r} - 1\right) R} \tag{11.2}$$

By combining Equations 11.1 and 11.2 we find a relationship between the force and voltages and V^+.

$$F = \frac{1}{\left(\frac{V^+}{V_r} - 1\right) R.a} - \frac{b}{a} \tag{11.3}$$

where $a = 0.079$, $b = -0.016$, R = 10 kΩ, and $V^+ = 5$ V.

The voltages and V^+ are transmitted to the data acquisition card and the force is calculated by the processing software from Equation 11.3 .

11.4 Shaft Position Sensor

A built-in linear potentiometer was used to find the position of the shaft. The relationship between resistance of the potentiometer and the length of the shaft is plotted in Figure 11.7.

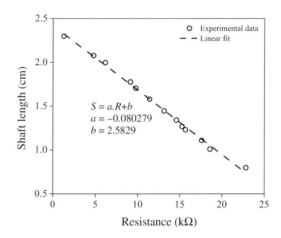

Figure 11.7   The relationship between shaft length and resistance

Using the linear relation of the shaft length and resistance, the position of the shaft could easily be determined by measuring the resistance of the potentiometer. To measure the resistance, a circuit, similar to that shown in Figure 11.6, was used in which the only difference was that the FSR was replaced by a potentiometer. The relationship between shaft length and the measured voltage V_s can be found in a similar way to Equation 11.3:

$$F = \frac{1}{\left(\frac{V^+}{V_s} - 1\right) R.a} - \frac{b}{a} \tag{11.4}$$

where $a = -0.080279$, $b = 2.5829$, $R = 10\ k\Omega$, and $V^+ = 5\ V$.

Therefore, the applied force F and the displacement of the shaft can be found using Equations 11.3 and 11.4.

11.5   Stress–Strain Curves

For this study, the mechanical properties of several materials were determined using mechanical compression tests for which the results are shown in Figure 11.8.

Each row in this figure shows an elastomeric material. The left column shows the stress–strain curve. The right column shows the real force–displacement relation for this same material. These force–displacement relations were used in this study to simulate the same elastomeric behavior with the actuator.

11.6   PID Controller

To replicate the behavior of elastomers, the linear actuator must follow the force–displacement relationship of each and, by so doing, simulate Young's modulus at

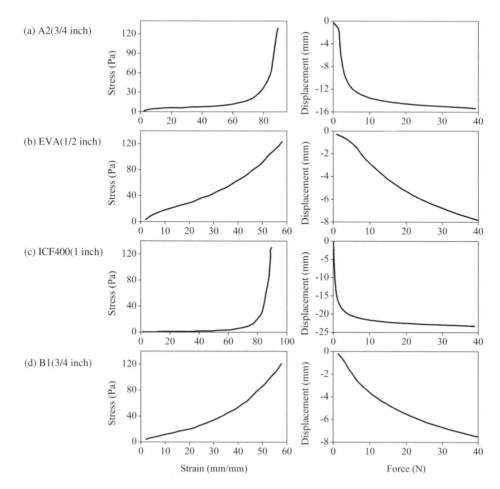

Figure 11.8 Mechanical compression test results for four elastomeric materials, left column: stress–strain relationship and right column: force–displacement relationship

each point of the curve. In order to accomplish this, a negative feedback closed-loop control system with feed-forward PID controller was used, which monitored the position of the shaft, compared its position against the desired position and prepared appropriate commands to the actuator in order to move the shaft accordingly.

In order to design a controller for the system, first a model for the linear actuator had to be found. Figure 11.9 shows the block diagram of the system.

For simplification, we neglected some blocks, such as the position sensor and the analog-to-digital and digital-to-analog converters, since these are linear and operate as simple unity gain. To find the transfer function for the linear actuator and the driver circuit, the PID controller was replaced by a unity gain device. Figure 11.10 shows the simplified block diagram of the linear actuator and the driver.

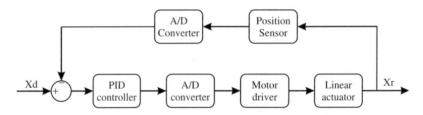

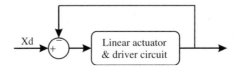

Figure 11.9 Complete block diagram of the system

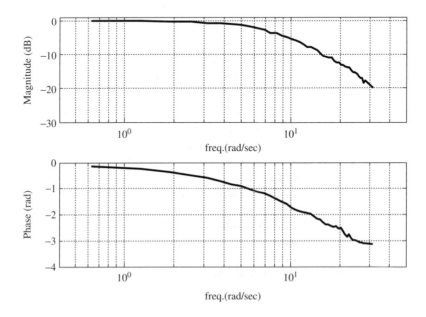

Figure 11.10 Simplified block diagram of the system

11.6.1 Linear Actuator Model

To identify the parameters in the transfer function of the actuator, a sinusoidal wave, with unity gain and variable frequency, was applied to the system. The frequency was changed over the range 0–5 Hz and the magnitude ratio between the output of actuator and input sinusoidal signal were registered, and also the time delays for each input. Then, using the experimental data, a Bode plot of the system (a graph of the transfer function of a linear, time-invariant system versus frequency, plotted with a log-frequency axis to show the system's frequency response), was traced out and shown in Figure 11.11.

Figure 11.11 Frequency response of the system

The parameters of the second-order system can be derived from the Bode plot. Knowing the general form of a second-order system:

$$G(s) = \frac{k\omega_n^2}{s^2 + 2\zeta\omega_n s + \omega_n^2} \tag{11.5}$$

where:

k is the system gain

ω_n is the system's natural frequency

ζ is the system's damping ratio.

The DC gain (k) of a system can be calculated from the magnitude of the Bode plot when $s = 0$ as $k = 10^{\frac{M(0)}{20}}$. The natural frequency of a second-order system occurs when the phase of the response is $-90°$ relative to the phase of the input ($\omega_n = \omega_{-90°}$). Where ($\omega_{-90°}$) is the frequency at which the phase plot is at $-90°$.

The damping ratio of a system can be found with the DC gain and the magnitude of the Bode plot when the phase plot is $-90°$:

$$\zeta = k / \left(2 \times 10^{\frac{M(-90°)}{20}} \right)$$

$$\begin{cases} DC\ gain : k = 10^{\frac{M(0)}{20}} \Rightarrow k = 1 \\ M(0) = 0 \end{cases} \tag{11.6}$$

$$\begin{cases} \omega_n = \omega_{-90°} \\ \omega_{-90°} = 9.3 \end{cases} \Rightarrow \omega_n = 9.3 \tag{11.7}$$

$$\begin{cases} \zeta = K / \left(2 \times 10^{\frac{M_{-90°}}{20}} \right) \\ M_{-90°} = 20\log(0.4) = -7.96\,\text{dB} \end{cases} \Rightarrow \zeta = \frac{1}{2 \times 10^{-0.4}} = 1.26 \tag{11.8}$$

Therefore the transfer function of the system is:

$$G(s) = \frac{86}{s^2 + 2(1.26)(9.3)s + 86} = \frac{86}{s^2 + 23.4s + 86} \tag{11.9}$$

Please note that $G(s)$ is the transfer function of the closed-loop system. To find the open loop transfer function, $H(s)$, consider Figure 11.12.

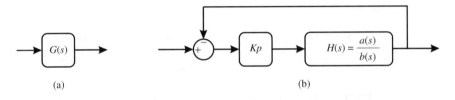

(a)   (b)

Figure 11.12   (a,b) Closed-loop system and its equivalent transfer function

In this figure, the closed-loop transfer function of the system is:

$$H(s) = \frac{a(s)}{b(s)} \Rightarrow G(s) = \frac{K_p.a(s)}{b(s) + K_p.a(s)} \tag{11.10}$$

In the experiments, we set $K_p = 1$. Therefore $H(s)$ can be easily derived from $G(s)$:

$$G(s) = \frac{a(s)}{b(s) + a(s)} = \frac{86}{s^2 + 23.4s + 86} \Rightarrow H(s) = \frac{a(s)}{b(s)} = \frac{86}{s^2 + 23.4s} \tag{11.11}$$

11.6.2 Verifying the Identification Results

To verify the identification results obtained in the frequency domain, a step input is applied to the system and the time response of the position of the linear actuator is checked. Figure 11.13 shows the response of the system to a step input.

In this figure, the overshoot can be calculated as:

$$OS = \frac{0.7}{5}100 = 14\% \tag{11.12}$$

And ζ can be calculated:

$$\zeta = \frac{-\ln(0.14)}{\sqrt{\pi^2 + (\ln(0.14))^2}} = 0.53 \tag{11.13}$$

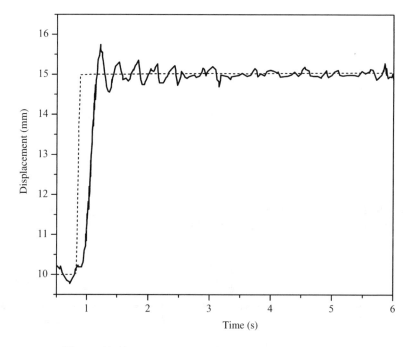

Figure 11.13   The response of the system to a step input

Using the following relationship, ω_d and ω_n can be found:

$$\omega_d = \frac{2\pi}{\Delta t}, \Delta t = \text{peak-to-peak time} \qquad (11.14)$$

$$\Delta t = 0.3 \, \text{seconds} \Rightarrow \omega_d = \frac{2\pi}{\Delta t} = \frac{2\pi}{0.3} = 6.67\pi \qquad (11.15)$$

$$\omega_n = \frac{\omega_d}{\sqrt{1 - \zeta^2}} = \frac{6.67\pi}{\sqrt{1 - 0.53^2}} = 24.7 \qquad (11.16)$$

$$G(s) = \frac{24.7^2}{s^2 + 2\,(0.53)\,(24.7)\,s + 24.7^2} = \frac{610}{s^2 + 26.2s + 610} \qquad (11.17)$$

Now the transfer function is calculated:

$$G(s) = \frac{K_p a(s)}{b\,(s) + K_p a(s)} \qquad (11.18)$$

$$K_p a\,(s) = 361, \, K_p = 7 \Rightarrow a\,(s) = 87 \qquad (11.19)$$

$$H(s) = \frac{a(s)}{b(s)} = \frac{87}{s^2 + 26.2s} \qquad (11.20)$$

Relationships 11.11 and 11.20 were derived from two different experiments and are very close to each other. The error that exists between the two relations is small and is considered to be normal due to the nonlinearities in the linear motor.

11.6.3 Design of the PID Controller

A PID controller can be designed using the following methods: the classical synthetic open-loop design, analytic state-space methods, and parameter optimizing [2].

11.6.3.1 Classical Synthetic Open-Loop Design

In this method, the designer starts with a simple controller and continues to add more terms to the controller action in the expectation that, after further additions, the controller will meet the design goals such that the controller is constructed in a compartmentalized manner in which a variety of tools are used at different stages [3–6].

11.6.3.2 Analytic State-Space Methods

This method, on the other hand, is based on an analytic solution of some optimal controller design problem, such as the linear quadratic Gaussian (LQG) problem [7–9].

11.6.3.3 Parameter Optimization Methods

Parameter optimization methods for linear time-invariant (LTI) feedback design starts with controller structures that are motivated by ideas from classical, modern, or other techniques. Fundamentally, however, LTI refers to the basic concept that whether we apply an input to the system now, or T seconds from now, the output will be identical except for a time delay of T seconds. What is generally meant by controller structure, however, is one in which the value of one or more parameters can be adjusted in any system model. One example is a PI (proportional-plus-integral) controller structure, which is a generic control loop feedback mechanism that is widely used in industrial control systems and is the most commonly used feedback controller. A PID controller calculates an 'error' value as the difference between a measured process variable and a desired set-point. The controller attempts to minimize the error by adjusting the process control inputs, which contain the coefficients (proportional gain) and K_i (integral gain), as well as the controller transfer function parameter $K_p + K_i/s$.

The next step in a parametric method is to select a quality system performance that will yield the most cost-effective and logical method, such as to acquire one from an already solved LQG problem. This is advantageous to the extent that the cost yielded by the structured controller by parameter search can then be compared to the absolute minimum achieved by any controller, which is analytically computable.

Another economic approach is to formulate a weighted sum or maximum of various performance indices, such as integrated square error in response to a step command, integrated magnitude of frequency response across some band where a disturbance is concentrated together with some indices representing actuator use. The idea of this is that the weights define the relative importance of different aspects of system performance. Finally, the designer may add explicit constraints on the values of the parameters, such as bounds on closed-loop pole locations and open-loop frequency responses.

After a controller structure is determined, together with cost functions and other constraints, the designer may need to rationalize. Many techniques for numerical solution of optimization problems resulting from control design problems have been proposed in control literature [10–15] although it is beyond the scope of this book to provide a comprehensive overview of these parameter optimization methods.

11.6.3.4 Controller Design

As previously mentioned, the position of the shaft, and the applied force by the finger to the shaft, should follow the desired force–position curve as closely as is possible. To minimize this tracking error, the optimum controller design method was used to find the best possible PID controller.

11.6.3.5 Solving the Problem in Simulink

The plant is a second-order linear system shown in Figure 11.14. The closed-loop system, including the actuator and the PID controller is shown in Figure 11.15.

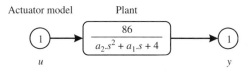

Figure 11.14   The plant is a second-order linear system

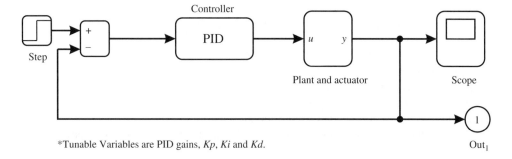

*Tunable Variables are PID gains, *Kp*, *Ki* and *Kd*.

Figure 11.15   The closed-loop system

The error is defined as the difference between step input and output. The cost function to be minimized (J) is the total square error from 0 to 100 seconds (see Equation 11.21).

$$J = \int_0^{100} \left[y(t) - u(t) \right]^2 dt \qquad (11.21)$$

The variables are the parameters of the PID controller, $J(K) = J(K_p, K_I, K_D)$. In discrete cases, Equation 11.21 is written as:

$$J\left(K_P, K_I, K_D\right) = \sum_{k=1}^{N} \left[y(k) - u(k) \right]^2 ; \quad k = 1, \ldots, N \qquad (11.22)$$

where:

$$N = \frac{100}{T_s} \qquad (11.23)$$

In which N is the total number of samples in 100 seconds and T_s is the sampling period.

Having the number of samples (N) and sampling period (T_s), the objective of the optimization problem is to find PID controller parameters for which $J(K)$ is minimized.

The Matlab routine '*lsqnonlin*' was used to perform least-squares fit on the tracking of the output. The tracking was performed via an M-file function '*tracklsq*,' which returns the error signal, the output was computed by calling '*sim*,' minus the input signal 1(unit step). The code for '*tracklsq*' is shown in Appendix 11.A.

Appendix 11.A

MATLAB CODE FOR PID OPTIMIZATION

```
function [Kp,Ki,Kd] = runm %runtracklsq
% RUNTRACKLSQ demonstrates using LSQNONLIN with Simulink.
% runm %model1 %optsim % Load the model
Pid0 = [0.91 0.105 0.1];% Set initial values
a1 = 21.3; a2 = 1; % Initialize plant variables in model
options = optimset('LargeScale', 'off', 'Display', 'iter', ...
'TolX',0.001 , 'TolFun', 0.001);
pid = lsqnonlin(@tracklsq, pid0, [], [],options);
Kp = pid(1); Ki = pid(2); Kd = pid(3);
function F = tracklsq(pid)
%Track the output of optsim to a signal of 1
%Variables a1 and a2 are needed by the model optsim.
% They are shared with RUNTRACKLSQ so do not need to be redefined here.
Kp = pid(1);
Ki = pid(2);
Kd = pid(3);
% Compute function value
simopt = simset('solver', 'ode14x', 'SrcWorkspace', 'Current');
% Initialize sim options
[tout,xout,yout] = sim('runm',[0 10],simopt);
F = yout-1;
end
Kp = pid(1)
Ki = pid(2)
Kd = pid(3)
%model1
end
```

The function '*runtracklsq*' sets up all the required values and then calls '*lsqnonlin*' with the objective function '*tracklsq*,' which is nested inside '*runtracklsq*.' The variable options passed to '*lsqnonlin*' define the criteria and display characteristics. In this case the medium-scale algorithm is used, and termination tolerances for the step and objective function are given on the order of 0.001.

To run the simulation in the model '*optsim*,' the variables a_1 and a_2 (a_1 and a_2 are variables in the Plant block) must all be defined. K_p, K_i, and K_d are the variables to be optimized. The function '*tracklsq*' is nested inside '*runtracklsq*' so that the variables a_1 and a_2 are shared between the two functions. The variables a_1 and a_2 are initialized in '*runtracklsq*.'

The objective function '*tracklsq*' must run the simulation. The simulation can be run either in the base workspace or the current workspace, that is, the workspace of the function calling '*sim*,' which in this case is the workspace of '*tracklsq*.' In this example, the '*simset*' command is used to tell '*sim*' to run the simulation in the current workspace by setting '*Src Workspace*' to '*Current*.' A solver for '*sim*' can be also chosen using the '*simset*' function. The simulation is performed using a fixed-step fifth-order method to 100 seconds.

When the simulation is completed, the variables and t_{out}, x_{out} and y_{out} are now in the current workspace (that is, the workspace of '*tracklsq*'). The '*Outport*' block in the block diagram model puts y_{out} into the current workspace at the end of the simulation. When you run '*runtracklsq*', the optimization gives the solution for the proportional, integral, and derivative (K_p, K_i, and K_d) gains of the controller after 20 function evaluations as K_p 0.9236, $K_i = 0.1654$, and $K_d = -0.0095$.

11.7   Processing Software

The processing software which was developed in the LabVIEW environment, receives the feedback signals from an FSR and a linear potentiometer (Figure 11.16). After filtering these signals and removing 60 Hz noise, the force (F) and shaft position (X_r) are calculated in separate modules using relationships 11.3 and 11.4. Then the calculated force is used to find the strain for the simulated material, which is saved in a lookup table (LUT).

After acquiring the dimensions of the material to be simulated, the stress is transformed to compression displacement (X_d) and then the shaft position, and the desired compression displacement (X_r and X_d), are fed to the PID controller. The output of the controller is finally transmitted to the output.

Figure 11.17 shows these parameters, which are plotted in LabVIEW. The force, shaft displacement, desired displacement, displacement error, and the output are sketched in real time. Figure 11.18 shows the programming environment and block diagram of the system.

11.8   Experiments

The following materials and elastomers: A2 ($^3/_4$ inch thick), EVA ($^1/_2$ inch), ICF400 (1 inch), A2-FR ($^3/_4$ inch), and B1 ($^3/_4$ inch), and a hard material (Plexiglas), were

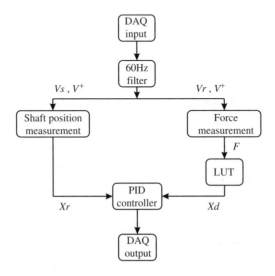

Figure 11.16   The flowchart of the processing software

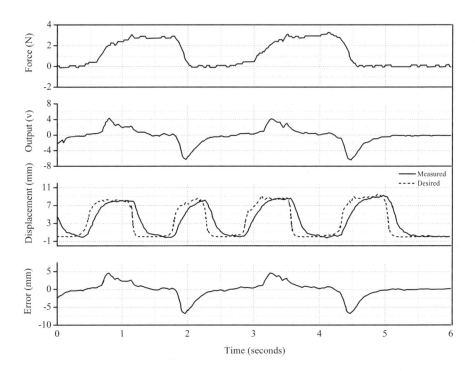

Figure 11.17   Measured and calculated signals plotted in real time

selected to be simulated by the actuator. The strain–stress relations for these materials are shown in Figure 11.19.

To compare the performance of the proposed tactile display, the force–position relation for each simulated material was saved, plotted, and compared to the force–position relation of the material. Each curve can be divided into two separate phases: loading and unloading. The loading phase starts when the finger makes contact with the shaft and applies force. As soon as the force is removed, the unloading phase starts.

The speed with which the material returns to its initial form during the unloading phase is dependent upon its composite structure and is, in any event, slower for softer materials. Figure 11.20 shows both phases in A2 ($^3/_4$ inch thick) and also the fact that, during the loading phase, a small error exists between the simulated curve and the real curve. An attempt was made to minimize the error by selecting an optimum controller although the error, as it transpired, was larger in the unloading phase, which did not, however, distort the perception of softness. The unloading phase, both in the original elastomer and linear actuator, started when the finger separated from the surface of the object, and from this it was established that the speed of unloading did not affect the sense of touch.

11.9   Results and Discussion

Figures 11.21–11.23 show the loading and unloading phases for three other elastomers (EVA $^1/_2$ inch, ICF400 1 inch, and B1 $^3/_4$ inch).

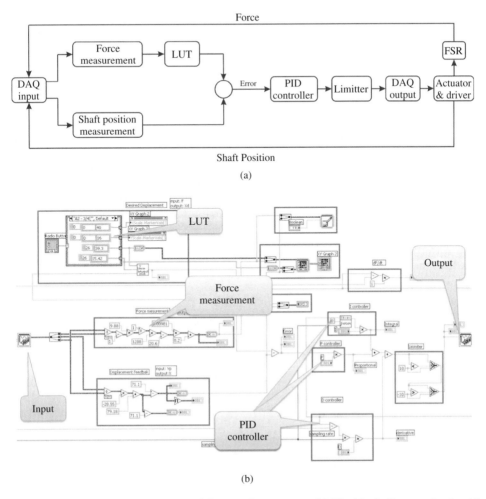

Figure 11.18 (a) The block diagram of the completer system. (b) The block diagram simulated in the LabVIEW programming environment. LUT: lookup table that contains the force–displacement data of the target material

The rate of applying the force is also important for the feel of softness. Since the rate of the force in the original force–deformation data collection was applied slowly, the error becomes negligible, as depicted in Figure 11.24. However, when the force is applied suddenly, the error becomes larger. Figure 11.24 shows the real and desired force–position curves for two different speeds. In the first instance, the force is applied slowly, whereas in the second case it is applied quickly. From this, it can be seen that the error in the first case is much less than in the second case. In this study, the force–displacement curve for the original elastomers was collected only at slow speed. However, in order to obtain a more comprehensive database on the elastomer, compression tests were conducted at different force or displacement rates and, based on the user speed at the tactile display side, the corresponding curve was used.

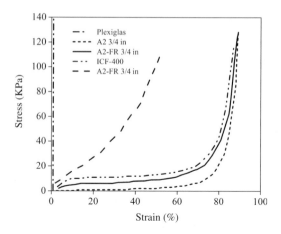

Figure 11.19 The stress−strain curves for different materials

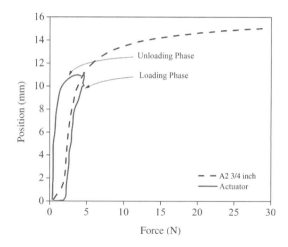

Figure 11.20 The force−position relation for A2 ($^3/_4$ inch). The dashed line is the original force−displacement curve and the continuous line is the one constructed in the tactile display

To evaluate the efficiency of the tactile display, another experiment was conducted in which an observer touched four elastomer materials with a finger and then the linear actuator. They then attempted to relate the real elastomers with the softness of the tactile display. For each person the experiment was repeated for the four simulated materials, as illustrated in Figure 11.25.

To remove the effect of visual feedback, the setup was covered so that the subjects were not able to see the different materials. In the first experiment, four materials (ICF400, A2, B1, and Plexiglas with Young's modulus = 1.2 GPa, were simulated by the tactile display. The subjects touched the materials and finally decided which of the objects had similar softness to the display. A total of 25 subjects participated in the experiment, all of whom successfully distinguished between the simulated elastomers (Table 11.2).

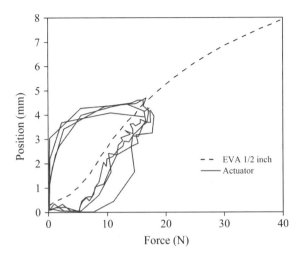

Figure 11.21   The position–force curve of EVA elastomer and its reconstructed equivalent utilizing tactile display

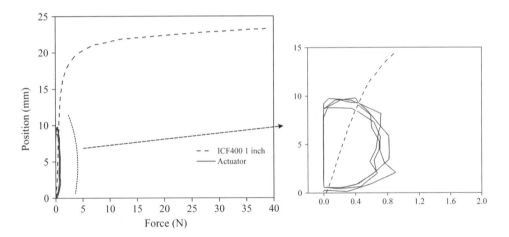

Figure 11.22   The position–force curve of ICF400 elastomer and its constructed equivalent

In the second experiment, two materials with very similar mechanical properties (H1N and ICF400) were simulated by the tactile display (Table 11.3). This time, the results of the simulation showed some discrepancies in recognition.

The reason is that this test required the subject to distinguish between two materials with similar mechanical properties, as shown in Figure 11.26, without visual feedback.

11.10   Summary and Conclusion

A system for displaying the softness of different materials, and which could potentially be used for telerobotic surgery and virtual reality, was proposed and tested and comprised a linear actuator, a force sensor, a data acquisition card, and processing software.

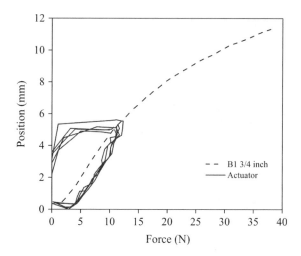

Figure 11.23   The force–position curve of B1 elastomer and its simulated equivalent

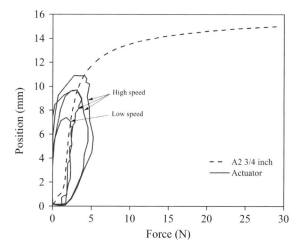

Figure 11.24   The effect of the force rate

Initially, the mechanical properties of the materials were characterized using mechanical compression tests. These data then were saved and used by processing software for reproducing the same properties in the tactile display. Having the dimension of the area over which the force was applied, these stress–strain data can be converted to force–compression which was used by the processing software to actuate the display. The data were saved in a lookup table in the processing software.

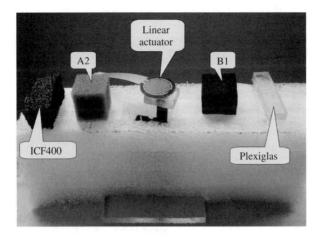

Figure 11.25 Experimental setup (See Plate 23)

Table 11.2 Results of the first human subject test

Simulated material	ICF400	A2	B1	Plexiglas
Recognized	25	25	25	25
Not recognized	0	0	0	0

Table 11.3 Results of the second human subject test

Material	A2-FR $\frac{3}{4}$ inch	A2 $\frac{3}{4}$ inch
Recognized	15	12
Not recognized	10	13

The processing software gathered information from the applied force to the shaft of a linear actuator and the position of the shaft. By having this information, and using the lookup table, the response of the material to the applied force was extracted from the table and used as the input for a PID controller, which prepared the necessary commands for the linear actuator to move the shaft.

Experiments on the tactile display were conducted by human subjects and the results were registered. Different materials were simulated by the tactile display and compared to the real objects. The results showed that the developed tactile display can replicate the softness of materials very closely and has proved to be a feasible technique with great potential for future use in robotic and minimally invasive surgery (MIS).

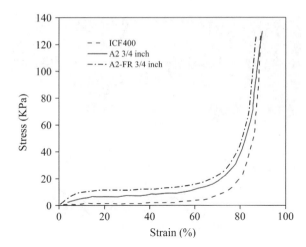

Figure 11.26 The material properties of the elastomers for the second experiment

References

1. Hannaford, B. and Okamura, A.M. (2008) Haptics, in *Springer Handbook of Robotics* (eds B. Siciliano and O. Khatib), Springer, Berlin, Heidelberg, pp. 719–739.
2. Boyd, S., Baratt, C., and Norman, S. (1990) Linear controller design: limits of performance via convex optimization. *Proceedings of the IEEE*, **78**, 529–574.
3. Bode, H.W. (1945) *Network Analysis and Feedback Amplifier Design*, Van Nostrand, New Your, NY.
4. Horowitz, I.M. (1963) *Synthesis of Feedback Systems*, Academic Press, New York, NY.
5. Horowitz, I.M. (1982) Quantitative feedback theory. *IEE Proceedings Part D*, **129**, 215–226.
6. MacFarlane, A.G.J. and Kouvaritakis, B. (1977) A design technique for linear multivariable feedback systems. *International Journal of Control*, **25**, 837–874.
7. Bryson, A.E. and Ho, Y.C. (1975) *Applied Optimal Control*, Hemisphere Publishing Co., New Your, NY.
8. Kwakemaak, H. and Sivan, R. (1972) *Linear Optimal Control Systems*, John Wiley & Sons, Inc., New Youk, NY.
9. Athans, M. and Falb, P. (1966) *Optimal Control*, McGraw-Hill, New York, NY.
10. Ly, U.-L., Bryson, A.E., and Cannon, R.H. (1985) Design of low-order compensators using parameter optimization, *Automatica*, **21**, 315–318.
11. Makila, P. and Toivonen, H. (1987) Computational methods for parametric LQ problems–a survey. *IEEE Transactions on Automatic Control*, **32**, 658–671.
12. Gangsaas, D., Bruce, K., Blight, J., and Uy-Loi, L. (1986) Application of modem synthesis to aircraft control: three case studies. *IEEE Transactions on Automatic Control*, **31**, 995–1014.
13. Polak, E., Mayne, D.Q., and Stimler, D.M. (1984) Control system design via semi-infinite optimization: a review. *Proceedings of the IEEE*, **72**, 1777–1794.
14. Polak, E., Siegel, P., Wuu, T. *et al.* (1982) Delight. MIMO: an interactive, optimization-based multivariable control system design package. *Control Systems Magazine, IEEE*, **2**, 9–14.
15. Fan, M.K.H., Li-Sheng, W., Koninckx, J., and Tits, A.L. (1989) Software package for optimization-based design with user-supplied simulators. *Control Systems Magazine, IEEE*, **9**, 66–71.

12

Clinical and Regulatory Challenges for Medical Devices

The term 'medical devices' covers a vast array of products, from simple tongue depressors to magnetic resonance equipment. The intended primary mode of action of a medical device on the human body, in contrast to that of medical products, is not metabolic, immunological, or pharmacological, although it may be assisted in its function by such means. With around 1.5 million different devices available, it is one of the fastest growing markets today. As a consequence, the regulatory approval and licensing of medical devices is becoming more and more challenging.

All medical devices, including minimally invasive surgical (MIS) and minimally invasive robotic surgical (MIRS) devices are governed by strict regulatory controls. Through various regulatory bodies, the US Department of Health and Human Services (HHS), and similar agencies in other countries, protect the public from a number of health risks and provides programs for public health and welfare. Together, these regulatory agencies protect and regulate public health at every level. For instance, the US Food and Drug Administration (FDA) is the federal regulatory agency responsible for public health through the regulation and supervision of food safety, tobacco products, dietary supplements, prescription, and over-the-counter pharmaceutical drugs (medications), vaccines, biopharmaceuticals, blood transfusions, medical devices, electromagnetic radiation emitting devices (ERED), veterinary products, cosmetics, food additives, product recalls, and restaurant inspection.

12.1 Clinical Issues

Medical devices remain distinct from drugs for regulatory purposes due to the fact that they operate via physical or mechanical means and are not dependent on metabolism to accomplish their primary intended effect. As defined in the federal Food, Drug, and Cosmetic

Tactile Sensing and Displays: Haptic Feedback for Minimally Invasive Surgery and Robotics, First Edition.
Saeed Sokhanvar, Javad Dargahi, Siamak Najarian, and Siamak Arbatani.
© 2013 John Wiley & Sons, Ltd. Published 2013 by John Wiley & Sons, Ltd.

(FD&C)[1] Act, the term medical device means an instrument, apparatus, implement, machine, contrivance, implant, *in vitro* reagent, or other similar or related item, intended for use in the diagnosis of disease or other conditions, or in the cure, mitigation, treatment, or prevention of disease, or intended to affect the structure or any function of the body, and which does not achieve its primary intended purposes through chemical action within, or on, the body.

Biocompatibility is a factor that must be considered for those medical devices which make contact with the body and adds even more complexity to the already challenging issues of medical devices. In general, a medical device consists of a variety of materials and is one of the factors that play an important role in the question of whether or not a medical device meets the requirements for human use (in other words whether it is 'biocompatible'). To consider the biocompatibility aspects, however, one must not only consider the applied materials, but the complete medical device as a whole and its intended use in the human body. The biocompatibility of a medical device depends on many factors, such as the time that it is exposed to the human body and the location in the body where it is applied or implanted. Also, the effect of a product and its materials on humans may vary, depending on its application, such as if it is noninvasive or invasive, and the frequency and duration with which it is used. Contact time and location in the human body varies from external to internal applications. Biocompatibility of these devices depends on the application within the human body. For instance, a material may be biocompatible for external application, but not internal use. Apart from the materials and its intended use, the biocompatibility of a medical device is also influenced by its production process, during which the quality of materials may decline and, as a consequence, the applied materials must be judged in their final state. Degradation can be caused by heat, temperature, or contact with other materials. Cleaning, packaging, and sterilization may alter the quality of materials. Therefore, the complete production process has to be taken into account in order to establish the biocompatibility of a medical device.

Society has benefited tremendously from the development and utilization of mechanical devices which are implanted inside the body to replace bones and joints, increase blood flow, and even measure blood chemistry. To further enhance the performance of these devices, the application of thin films to external surfaces is an ongoing research and development interest in many companies. Engineers have the choice of a variety of technologies to apply liquid coatings to these often complex surfaces, ranging from vacuum technology to direct liquid application. Although most of the materials used in medical devices have been investigated at the bulk scale for their biocompatibility, the toxicity and hemocompatibility of certain materials is still to be determined and more rigorous research studies need to be undertaken. Thin films are much more convenient and preferred to coatings. As the properties of thin films depend heavily on their deposition process, further research needs to be carried out on the behavior of thin films compared to

---

[1] The United States Federal Food, Drug, and Cosmetic Act (abbreviated as FFDCA, FDCA, or FD&C), is a set of laws passed by Congress in 1938 giving authority to the US Food and Drug Administration (FDA) to oversee the safety of food, drugs, and cosmetics. A principal author of this law was Royal S. Copeland, a three-term US Senator from New York [1]. In 1968, the Electronic Product Radiation Control Provisions were added to the FD&C. Also in that year the FDA formed the Drug Efficacy Study Implementation (DESI) to incorporate into FD&C regulations the recommendations from a National Academy of Sciences investigation of the effectiveness of previously marketed drugs [2]. The act has been amended many times, most recently to add requirements about bioterrorism preparations.

their bulk form. Preliminary results indicate that there are no cellular toxicity effects and the risk of clotting is only slightly increased with these conventional materials used in medical devices [3]. One approach, in using noncompatible medical devices *in vivo*, relies on isolating medical devices from the body by packaging them in biocompatible polymers although this can add to the size of medical devices and reduce their accuracy. To overcome these issues, nanoparticle coatings and biocompatible polymer micromachining need to be investigated further.

Packaged medical devices must be able to survive the sterilization procedures used in the surgical environment and preferably be able to withstand exposure to high temperatures and moisture in autoclaves and steam sterilizers. Alternative sterilization methods include ethylene oxide (ETO) and irradiation. ETO is a harsh organic solvent and packaging must be made of a compatible material. Medical devices are inherently radiation hardened, but their associated electronics are not, so they must be specially designed using radiation-hardened IC processes and packaging.

Since it is important to ensure that medical devices and products remain sterile in the packaging throughout distribution, in order to allow their immediate use by physicians, medical device packaging is highly regulated. A series of special packaging tests is used to measure its ability to maintain sterility. Some of the relevant standards include: ASTM F1585-Guide for Integrity Testing of Porous Barrier Medical Packages [4], ASTM F2097-Standard Guide for Design and Evaluation of Primary Flexible Packaging for Medical Products [5], EN 868 Packaging materials and systems for medical devices which are to be sterilized [6], General requirements and test methods, and ISO 11607 Packaging for Terminally Sterilized Medical Devices [7].

Designing well-controlled prospective clinical trials of medical devices presents unique challenges that differ from those faced in the study of pharmaceuticals. For example, clinical effects observed in medical device studies are influenced not only by the product under evaluation and the patients themselves, but also by the skill and discretion of the user, who is typically a health care professional, but who may also be the patient. The impact of this parameter (the medical device user) is a variable unique to medical device studies and can be responsible for the greatest degree of variability in clinical outcomes. One critical variable that requires attention in designing a clinical study is being aware of the extent to which the user is able to influence and control the device performance [8]. As indicated, hemocompatibility and biocompatibility, being resistant to sterilization procedures and user parameters, currently are the most important clinical issues for medical devices.

12.2 Regulatory Issues

The furtherance and control of quality is a major issue in health care. It is also one of the aspects of health technology assessment which, in large part, consists of the quality, safety, and effectiveness of medical devices that are employed in diagnosis and treatment. Medical devices, of which surgical tools and telesurgical systems are subsets, are subject to many regulatory controls. The FDA and European Community (EC) determine whether a product is fit for sale in the United States and Europe, respectively. The regulation of medical devices is a vast and rapidly evolving field that is often complicated by legal technicalities. For example, legal terms and their meanings are sometimes nonuniform even within one regulatory system.

12.2.1 Medical Product Jurisdiction

When preparing a regulatory strategy for a product or technology, it is important to first establish if the product is a device, a drug, or a biologic, and then to determine which regulations apply. Two factors must be considered when making this distinction. First, the manufacturer or supplier must clearly state and indicate the product's purpose and, secondly, the exact way in which it is to be applied and/or used. From this, it is possible to determine if the product is chemical/metabolism-altering (a drug) or physical in nature (a device). For example, if an alginate wound dressing contains an antibacterial agent whose sole purpose is to act as a barrier *between* the wound and the environment, it would be classified as a device, since it is performing a physical function. On the other hand, if the indication for use is to deliver the antibacterial agent (chemical) *onto* the wound itself in order to treat an existing infection, then the alginate dressing is considered to be a drug. In order to make this determination, one must carefully review the definition of a medical device contained in the 1976 Medical Device Amendments of the Food, Drug, and Cosmetic Act [9]:

> An instrument, apparatus, implement, machine, contrivance, implant, *in vitro* reagent, or other similar or related article, including any component, part, or accessory, which is:
>
> 1. Recognized in the official National Formulary, or the United States Pharmacopeia (USP), or any supplement to them,
> 2. Intended for use in the diagnosis of disease or other conditions, or in the cure, mitigation, treatment, or prevention of disease, in man or other animals, or
> 3. Intended to affect the structure or any function of the body of man or other animals, and which does not achieve its primary intended purposes through chemical action within or on the body of man or other animals and which is not dependent upon being metabolized for the achievement of any of its principal intended purposes.

12.2.2 Types of Medical Devices

There is a wide variety of medical devices in use today. They range from room-sized imaging systems that weigh several tons, to ophthalmic implants that are less than 2 mm long and weigh only a few grams. Most *in vitro* diagnostic products (blood and urine tests) are also regulated as medical devices. Table 12.1 below describes most devices using two of their characteristics.

The left column shows the device function, and the right column shows its form. For example, a lithotripter that uses sound waves to break up kidney stones would be considered a durable therapeutic device; a pacemaker would be considered an implantable therapeutic device, and so on. Issues such as reuse, shelf life, and device tracking impact different types of devices in different ways.

Table 12.1 Medical device types [10]

Function	Form
Therapeutic	Durable
Monitoring	Implantable
Diagnostic	Disposable

12.2.3 Medical Device Classification

Section 510(k) of the Food, Drug, and Cosmetic Act requires device manufacturers to notify the FDA of their intent to market a medical device at least 90 days in advance. This is known as premarket notification — also called PMN or 510(k). Once a determination has been made that a product is a medical device, the next issue that must be addressed is its classification. In other words 'What kind of submission do I need to commercialize this device? Is it exempt from 510(k) notification requirements, subject to those requirements, or must we file a premarket approval application (PMA)?' In order to answer these questions, we first need to know the class of the device.

Regulatory control increases from Class I to Class III. Class I devices are the simplest and are only subject to general controls since they pose the fewest risks. They must, however, follow the regulations and have accurate labels and labeling. Furthermore, since the manufacturer must register and list its devices with the FDA, it cannot, therefore, be a banned device or be subject to a recall or similar action. Most Class I devices are exempt from PMN requirements (510(k) [2]), and some are also exempt from compliance with the quality system regulations (QSR). Examples of Class I devices include toothbrushes, oxygen masks, and irrigating syringes. The FDA estimates that approximately half of the medical devices it regulates are Class I devices.

If Class I controls are not, by themselves, sufficient to provide reasonable assurance of safety and effectiveness, the device is placed in the Class II category, in which special controls are imposed due to the moderate risk involved. In order to market a Class II device in the United States, the manufacturer must obtain clearance of a 510(k) PMN prior to commercialization. The purpose of this is to demonstrate that the new device is substantially equivalent to either another device that has already gone through the 510(k) process or one that was on the market before the Medical Device Amendments were signed on 28 May 1976. Class II devices are subject to the special controls as outlined in the Office of Device Evaluation (ODE) guidance documents, FDA-accepted international standards, and the QSR. These controls include a review of performance standards, post-market surveillance, patient registries, and submission of guidelines (including clinical data in PMN). Devices that fall into this category include ultrasound imaging systems, Holter cardiac monitors, pregnancy test kits, and central line catheters. The FDA estimates that slightly less than half of the medical devices it regulates are Class II devices, of which approximately 3200 are cleared each year for use on the US market [11].

If general controls and special controls (Classes I and II) are still insufficient to ensure the safety and efficacy of the device, an application for Class III (premarket) approval

---

[2] These notifications, usually called 510(k)s because they are regulated by section 510(k) of the Medical Device Amendments, are reviewed by the FDA.

must be made. This includes reports of investigations to show whether or not such a device is safe and effective, together with a statement of the components, ingredients, properties, and principle(s) of operation. A description is also required on the methods, facilities, and control for the manufacture, processing, packing, and installation of the device. In addition, reference should be made to any relevant performance standards and the ability of the device to meet them. Usually, it is also required to provide a sample of the device and its components, together with a sample of proposed labeling and certification related to clinical trials, and any other relevant information the FDA may require. Most Class III devices require PMA approval prior to marketing in the United States. These are devices that are not substantially equivalent to any Class II device and are usually technologically innovative. A manufacturer who wishes to have a device reclassified to a lower class must convince the FDA that the less stringent class requirements will be sufficient to provide reasonable assurance of safety and effectiveness. There are still a small number of Class III pre-amendments 510(k) devices; however, the FDA has been working diligently to either reclassify them to Class II or, if their risk profile does not justify this, call for PMAs. There were 39 PMAs approved in 2006 [12].

12.2.4   Determining Device Classification

If the product in development is similar to other medical devices already in use on the US market then, with respect to its indication for use and its technological characteristics, the regulations need to be searched in order to determine its classification. The 21 Codes of Federal Regulation (CFR) 862–892 contains descriptions of a wide variety of medical devices arranged by medical practice area. The classifications and exemptions from 510(k) or QSR, if any, are listed in this section of the regulations. The classification database on the Center for Devices and Radiological Health (CDRH) web site can also be a useful tool for determining device classification (Table 12.2).

If a description in the CFR is consistent with the characteristics of the new device, then the device classification listed in that section of the CFR should apply. Precedents can be identified in another manner as well. If one is aware of other competing devices that are already on the market, one can search the 510(k) or PMA databases [11], [12] within the CDRH web site for those products, and determine how they were classified.

When there is no obvious precedent to follow, it can be difficult to determine the appropriate device classification. Under Section 513(g) of the Food, Drug, and Cosmetic Act (the Act), device developers can request a classification decision from the FDA for a new or modified device for which there is no clear classification already available. The FDA's written guidance, as a result of a 513(g) submission, can be valuable in making decisions about the appropriate regulatory strategy for product commercialization. This guidance is also sometimes useful in discussions with potential distributors, marketing partners, or investors. In addition, all of the elements of 513(g) may be incorporated into the 510 (k) PMN should the FDA classify the product as a Class II device. Device developers can obtain a formal classification decision using the 513(g) request for classification process. The sponsor submits a brief document to the ODE describing the device, how it works, materials used, and similar devices, if any. The indication for use and draft labeling is also included along with a suggested classification and supporting rationale.

Table 12.2   Medical device classification [10]

Device classification panel or specialty group	21 CFR* part
Anesthesiology	868
Cardiovascular	870
Clinical chemistry and clinical toxicology	862
Dental	872
Ear, nose, and throat	874
Hematology and pathology	864
Immunology and microbiology	866
Gastroenterology and urology	876
General and plastic surgery	878
General hospital and personal use	880
Neurology	882
Obstetrical and gynecological	884
Ophthalmic	886
Orthopedic	888
Physical medicine	890
Radiology	892

*CFR, Code of Federal Regulations.

12.3   Medical Device Approval Process

Once the classification of the device has been determined, it is then necessary for the manufacturer to determine which regulatory control (Class I, II, or III) will apply to their medical device and hence its pathway onto the US market. For example, a device such as software that analyzes magnetic resonance imaging (MRI) images would be designated as a Class II 510(k) product if its purpose was only to measure the size or volume of anatomical structures. On the other hand, if the purpose of the software was to detect abnormalities or provide diagnostic information, it would be considered a Class III PMA device. This example highlights how critical it is to provide an exact and detailed description as to what the device is designed to accomplish. A device developer may choose to 'start small' and begin interactions with the FDA by producing a device that requires only the more simplistic aspects of 510(k) to be considered and then, after gaining experience, move on to the more challenging PMA once a revenue stream has been established. Generally, both industry and the FDA would prefer to review medical devices as 510(k)s since it provides the manufacturer with timely reviews and conserves reviewing resources for the FDA. So, when speed to market is the prime consideration, one always attempts to follow the 510(k) path, although, even within this, there are a number of branches. If the FDA-recognized standards apply to the new device, the sponsor may choose to submit an abbreviated 510(k) or a traditional 510(k). The review time is the same, but, instead of containing the complete test reports, an abbreviated 510(k) contains a summary of the test results and a list of the recognized standards followed during device testing for which, therefore, only a smaller submission is needed. Of course, a sponsor may choose an alternative test method, in which case the test protocol would need to be included in a traditional 510(k). In some cases, device developers with sufficient resources may choose to propose the more complex PMA designation and run clinical trials due to either unclear device classification or being unable to fully vouch for their product. This strategy, which presents a barrier for other less well-funded organizations, is often called creation of a 'regulatory patent.' Another consideration when deciding on a regulatory path is user fees. Since October 2002, the ODE has been authorized to charge fees for reviewing 510(k)s, PMAs, and PMA supplements.

12.3.1 Design Controls

Once the product definition and regulatory strategy have been prepared, Classes II and III device developers must work to comply with the design control provisions of the QSR (21 CFR 820) as the device development process moves forward. The QSR is the medical device equivalent of the pharmaceutical current good manufacturing practices (cGMPs). The QSR, unlike cGMPs, also regulates the device development process via its design control provisions (21 CFR 820.30) and it is the purpose of this section to describe the device developer's obligations.

Other sections of the QSR are discussed in Section 12.3.4. It should be noted that the preamble to the QSR [14] states that they only regulate development activities and not research activities. Although the regulation does not provide guidance for distinguishing between these two activities, the preamble does add that 'The design control requirements are not intended to apply to the development of concepts and feasibility studies. However, once it decides that a design will be developed, a plan must be established.'

Most device developers categorize research as being investigations into general technology, and development as being the formulation of any resulting product or application. For example, if a device developer creates a new laser technology, that effort would be considered research. Once the developer begins to apply that technology to a particular device with specific indications for use and user requirements, then they have begun the development phase and design controls must be applied. A device developer's design control standard operating procedures (SOPs) should clearly describe the point in the development process when design controls apply, and that definition should be consistently followed for all design projects.

There are components of design controls that stretch from planning for the development effort through design transfer (from development to manufacturing) and maintenance of existing designs. These controls apply to all Classes II and III medical devices and a small number of Class I devices. The purpose of these controls is to ensure that devices are developed in a rational manner and in compliance with the manufacturer's existing design control SOPs. Figure 12.1 represents the selected pathways for marketing medical devices in the United States.

12.3.2 The 510 (K) Premarket Notifications

More than 3000 medical devices are cleared for use on the US market every year through the 510(k) PMN process. This represents approximately half the new devices that appear on the US market in a given year. The 510(k) process is relatively rapid, flexible, and adaptable to many different device types and risk levels. The goal of the 510(k) process is demonstration of substantial equivalence to a device that was on the US market prior to

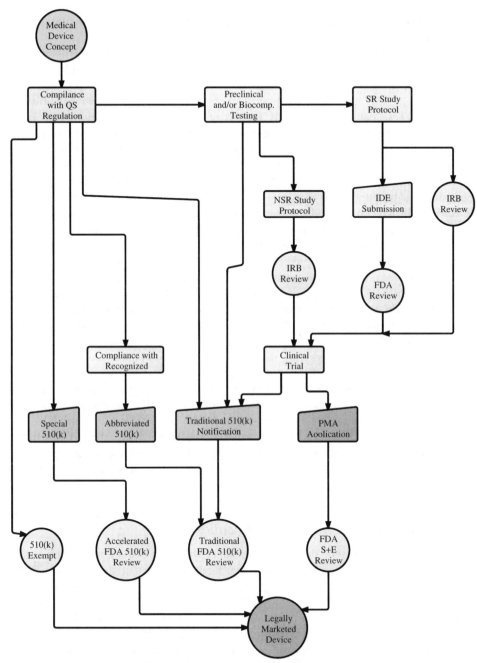

Abbreviations: SR, Significant Risk; NSR, Non0significant Risk; IDE, Investigational Device
Exemption; IRB, Institutional Review Board;

Figure 12.1   Selected pathways for marketing medical devices in the United States [10] (© Taylor & Francis Group)

28 May 1976 or to a device that has already gone through the 510(k) clearance process. Devices that have already successfully gone through the 510(k) process are described as '510(k) cleared,' whereas PMA devices are 'approved.'

Under section 510(k), a previously cleared device included for comparison purposes is called a predicate device and may contain multiple predicate devices that address various features of the device. The device designers should provide regulatory personnel with assistance by identifying key technological characteristics that demonstrate substantial equivalence. These data should already be part of the design inputs required as part of design controls, although, generally, little manufacturing data are included in 510(k). Sterile devices include information on the sterilization process, including validation activities and the sterilization assurance level. *In vitro* diagnostic products will frequently include data on the production of key reagents such as antibodies or nucleic acid probes. The other part of substantial equivalence relates to the indication for use since, frequently, one medical device can be used for many indications in a variety of medical specialties. When new indications are added, these must be cleared in a traditional or abbreviated 510(k) in which a predicate device with the same indication for use must be cited.

When searching for potential predicate devices, several information sources are useful. Two FDA databases, the 510(k) database [11], and the classification database [13] can be very helpful. The 510(k) database is especially useful when one knows either the name of potential predicate devices or the manufacturer of the device. The classification database can be used to identify a particular device type and its corresponding product code. One can then transfer the product code to the 510(k) database and generate a listing of all similar devices. Sales and marketing staff, together with competitor web sites, are also excellent sources of predicate device information [10].

As a periodical assessment of the effectiveness of existing programs, the FDA launched a comprehensive assessment of the 510(k) process in 2009. The combined result of the FDA internal assessment and an evaluation done by the Institute of Medicine (IOM) was a report with 55 recommendations to improve the process. Currently the FDA intends to initiate 25 new actions to implement 47 of the 55 recommendations, including development of new guidance on staff training. The new 510(k) action plan will be implemented to improve product safety.

12.3.3   The Premarket Approval Application

PMAs are necessary when the device the developer wishes to market is innovative in the United States and is not substantially equivalent to any other device that has been cleared through the 510(k) process. By necessity, the PMA process is much more stringent than the 510(k) process since it must demonstrate that the device is safe as well as effective, and typical review times are approximately one year. Unlike most 510(k) processes, a detailed manufacturing section describing the methods for building and testing the device must be included. Prior to final approval of the PMA, the CDRH office of compliance must review and approve the results of a preapproval inspection of the device manufacturing and development facilities. The sponsor of the clinical trial, and two or three of the clinical investigation sites, are also often subject to CDRH bioresearch monitoring (BIMO) inspections to confirm compliance with relevant sections of 21 CFR 812. Finally, the postmarket requirements of a PMA are considerably more complex than those related to a 510(k). Specifically, a PMA annual report must be filed with the ODE each year and changes in labeling, materials, manufacturing, quality methods, and specifications as well as changes in manufacturing location must all be reported to, and approved by, the ODE in advance. This is done through the PMA supplement process.

12.3.3.1   The PMA Process

PMAs are large and complex documents, often greater than 2000 pages. It can frequently take several years to obtain all the preclinical, clinical, and manufacturing data necessary for the PMA. It is essential that the PMA preparation effort be well planned, with good coordination between all functional areas involved in the development process and advance research, before a regulatory strategy is prepared and it should also include a wide variety of sources. Shortly after a PMA device is approved, the approval letter, summary of safety and effectiveness, and official labeling are placed on the CDRH web site. These documents provide greater technical and regulatory detail than a 510(k) summary. The PMA submission itself is not available via the Freedom of Information process.

Once the indication for use and the device description have been established, it is important to confirm the key elements of the development plan with the appropriate reviewing branch within the ODE. The device developer may choose to obtain this information via an informal telephone call, an informal pre-IDE (Investigational Device Exemption) meeting, a formal designation meeting or a formal agreement meeting. Less formal meetings, while they do not generate binding agreements, can encourage very productive technical exchanges. The choice of meeting type involves balancing business, regulatory, and clinical needs.

A PMA development plan should be executed when it has been established and reviewed by the ODE. Generally, multiple activities, such as manufacturing development, validation, preclinical functional testing, biocompatibility testing, and clinical testing proceed along parallel and often simultaneous lines. In some cases, it may be clear during the planning phase that some data, such as information that pertains to the manufacturing process or preclinical testing data, may be available long before the clinical trial has ended, in which case a process called a modular PMA may be advantageous. This consists of submitting completed pieces of the PMA to the ODE rather than sending all the data at the very end. For this, a sponsor must submit a shell or outline of the PMA and obtain approval from the ODE. The shell, which describes the contents of each module, is submitted to the ODE for independent review. Once the review of a module has been successfully completed, the ODE sends the sponsor a letter stating that the module is 'locked' and will not be reopened unless some portion of data already submitted changes in later stages of the development process. When the last module is submitted, the ODE considers the PMA complete [10].

12.3.4   The Quality System Regulation

The QSR regulates both the device development and the manufacturing process for all Class II and Class III devices from the beginning of the development phase until the device is no longer supported by the manufacturer. It covers the manufacturing process for a number of Class I devices, but does not cover the research process for any medical devices. The goal of the QSR is to create a self-correcting system that reliably produces robust device designs and production methods, ensuring that devices perform in a manner consistent with their intended use. In many ways, the QSR has evolved into the glue that holds the medical device regulatory process together from development through end of use. Once a device is marketed, the corrective and preventive action (CAPA) provisions of the QSR are closely related to compliance with the Medical Device Reporting (MDR) regulations. An additional advantage of the QSR is that it follows the philosophy of the international medical device standard, ISO 13485, which helps to enable device companies that sell their product internationally to maintain common systems for most design- and production-related activities. The manufacturing and quality processes also require specific evaluations and procedures, all of which must be documented. Frequently, the FDA field investigators will follow the quality system inspection techniques (QSITs) approach when inspecting a device facility [10].

12.4 FDA Clearance of Robotic Surgery Systems

Following the success of the PUMA 560 robotic surgical arm used in nonlaparoscopic delicate neurosurgical biopsy (1985), and later in laparoscopic cholecystectomy and transurethral resection (1987), the first FDA approved (cleared in 1994) surgical robot AESOP (Automated Endoscopic System for Optimal Positioning) was introduced by Computer Motion, Inc. in 1990. In 1995, the Zeus robotic surgery system was demonstrated by Computer Motion and tested on animals. By 2000, the Zeus was equipped to hold 28 different surgical instruments, and in 2001 it received FDA approval.

In 1999, Intuitive Surgical Devices, Inc. commercialized the 'Leonardo' and 'Mona' prototypes of their robotic surgical system, calling it the da Vinci System, and began marketing it in Europe while awaiting FDA approval in United States. In 2000, the da Vinci surgical system was approved by the FDA as the first robotic system for general laparoscopic surgery. FDA clearance was based on a review of clinical studies of safety and effectiveness submitted by the manufacturer and on the recommendation of the General and Plastic Surgical Devices Panel of the FDA's Medical Devices Advisory Committee.

References

1. Potter, R.J. (1967) Royal Samuel Copeland, 1868–1938: a physician in politics. PhD, Western Reserve University, Cleveland, OH.
2. Food and Drug Administration (FDA) (2011) Compliance Policy Guides, CPG Sec. 440.100 Marketed New Drugs without Approved NDAs and ANDAs.
3. Roy, S. (2002) BioMEMS for minimally invasive medical procedures. Presented at the The BioMEMS 2002 Conference, Boston, MA.
4. ASTM International (2000) ASTM Standard F1585. *Guide for Integrity Testing of Porous Barrier Medical Packages*, West Conshohocken, PA (Withdrawn 2006).
5. ASTM International (2010) ASTM Standard F2097. *Guide for Design and Evaluation of Primary Flexible Packaging for Medical Products*, West Conshohocken, PA.
6. British Standards Institution (2009) BS EN 868–5. *Packaging for Terminally Sterilized Medical Devices. Sealable Pouches and Reels of Porous and Plastic Film Construction. Requirements and Test Methods*.
7. International Organization for Standardization (ISO) (2006) ISO 11607. *Packaging for Terminally Sterilized Medical Devices*.

8. Becker, K.M. and Whyte, J.J. (2006) *Clinical Evaluation of Medical Devices*, Humana Press, Totowa, NJ.

9. Food and Drug Administration (FDA) (1976) Medical Device Amendments of the Food, Drug, and Cosmetic Act, Definition 201(h).

10. Pisano, D.J. and Mantus, D.S. (2008) *FDA Regulatory Affairs: A Guide for Prescription Drugs, Medical Devices, and Biologics*, 2nd edn, Informa Healthcare, New York, NY.

11. Food and Drug Administration (FDA) (2011) 510(k) Database.

12. Food and Drug Administration (FDA) (2011) PMA Database.

13. Food and Drug Administration (1996) Final rule. Medical devices; current good manufacturing practice (CGMP) final rule; quality system regulation. *Federal Register*, **61**(195), 52602–52662.

14. Food and Drug Administration (FDA) (2011) Product Classification Database.

Index

Tactile Sensing and Displays: Haptic Feedback for Minimally Invasive Surgery and Robotics, First Edition.
Saeed Sokhanvar, Javad Dargahi, Siamak Najarian, and Siamak Arbatani.
© 2013 John Wiley & Sons, Ltd. Published 2013 by John Wiley & Sons, Ltd.